P9-AOQ-710

NAFTA AND THE ENVIRONMENT

2044 030 041 124

WITHDRAWN

NAFTA AND THE ENVIRONMENT

Terry L. Anderson, Editor

Pacific Research Institute for Public Policy, San Francisco

HF
1746
.N33
1993
C.2

Copyright © 1993 The Fraser Institute. All rights reserved. No part of this publication may be reproduced, stored in a retrieval system, or transmitted in any form or by any means, electronic, mechanical, photocopy, recording, or otherwise, without the prior written consent of the publisher.

RECEIVED

MAR 2 5 1995

Kennedy School Library

ISBN 0-936488-73-5

Printed in the United States of America
1   2   3   4   5   6   7   8   9   10

Pacific Research Institute for Public Policy
177 Post Street, Suite 500
San Francisco, CA 94108
(415) 989-0833

Library of Congress Cataloging-in-Publication Data

NAFTA and the environment / Terry L. Anderson, editor.
    p.   cm.
ISBN 0-936488-73-5 (pbk.)
1. Free trade—North America. 2. Free trade—Environmental aspects—North America. 3. Canada. Treaties, etc. 1992 Oct. 7.
I. Anderson, Terry Lee, [date]
HF1746.N33   1993   93-1031
363.7'0097—dc20   CIP

Cover Design: *Praxis*
Text Design: *Dianna Rienhart*
Printing and Binding: *Data Reproductions Corporation*

Contents

Acknowledgments

Obviously, a volume like this is a combined effort of the authors and the editor, but there were many other producers. The Fraser Institute of Vancouver, British Columbia, under the leadership of Michael Walker, provided the impetus for the project. Had he not obtained funding from the Lilly Endowment, the first words would not have been written.

Bringing the volume to the camera-ready stage required the diligent efforts of the staff at the Political Economy Research Center, Monica Lane Guenther, Dianna Rienhart, and Pam Malyurek. I only hope they never figure out that I am the cog in the wheel that is dispensable.

Finally, Sally Pipes, President of the Pacific Research Institute, deserves special thanks for agreeing to publish the volume and Kay Mikel for overseeing the final stages of publication.

Terry L. Anderson
Bozeman, Montana
February 1993

Introduction

Terry L. Anderson

With the European Community setting the precedent for trade liberalization among a "community" of trading partners, groups of countries around the world have followed suit. Thus, today virtually every continent has its existing or proposed alliances for trade. Overarching many of these has been the General Agreement on Tariffs and Trade (GATT) which creates multilateral trade agreements across the oceans.

The tremendous potential for trade liberalization is exemplified by the North American Free Trade Agreement or the NAFTA. This agreement would combine the world's largest, eighth-largest, and thirteenth-largest economies. A unified North American market would constitute 360 million people with a total purchasing power of $6.2 trillion. (Note that throughout this book where figures are given in dollars, they refer to U.S. dollars at 1992 exchange rates, unless otherwise specified.) While the European Community has more people, it embodies less income and spending power. Trade between the United States and Mexico alone in 1992 exceeded $70 billion, almost triple what it was five years earlier.

The positive impacts of trade liberalization notwithstanding, arriving at such agreements has not been easy. It took decades to establish the European Community, and parties to the GATT have been at the negotiating table through many rounds. If the trade liberalization is such a good thing, why the difficulty in reaching agreements? Isn't trade a positive-sum game? Aren't there enough gains from trade to make all parties happy?

The simple answer is that the distribution of gains from trade is often the "trump card." It is true that trade generates a *net* increase in jobs, income, and wealth, but the "net" creates the sticking point. It is easy to say that those whose jobs are displaced by foreign competition have additional opportunities to find new jobs, but for the displaced worker, finding the new job has its costs. Free trade does little for the personal financial statement of the stockholder whose company cannot compete with foreign products even if the trade does increase "the wealth of nations." And for the person who retains his job but breathes dirtier air, trade liberalization may not be an improvement.

Hence nearly all trade agreements face at least two sticking points: job protection and the environment. The former is nothing new to the trade scene. Indeed, British mercantilism, which stimulated much of Adam Smith's political economy, was a protectionist scheme supplied by the Crown to the guilds, unions, and monopoly companies in return for tariff revenues. But protectionist schemes are sometimes too blatant and therefore arise under new and surprising guises.

With the recent round of trade agreements and especially the NAFTA, environmentalists have joined the ranks of those concerned with the implications of trade liberalization. Interestingly, much of the debate and discussion regarding the NAFTA has not centered on the economic impacts. Rather, it has centered on the potential damage to the environment, the threat to native culture, and the implications for democracy and human rights. The introduction of these issues into the trade debate has created some strange bedfellows, to wit the "greening" of organized labor as it has joined forces with environmentalists in opposition to freer trade.

At the same time that trade pacts have been on the negotiating table, the concept of "free market environmentalism" has been increasingly recognized as an alternative to command-and-control environmental policy. Two themes particularly relevant to the trade liberalization debate emerge from free market environmentalism. The first is that economic growth and environmental quality generally go hand-in-hand. The logic of the argument is simple; if environmental quality is typical of most goods, consumers will want more of it as their incomes rise. They may choose more green space by purchasing larger yards or gardens; they may join the throngs of "eco-tourists" heading to the rain forest; or they may demand that their government enforce stricter environmental standards on those who dispose of their waste into the common environment. And with empirical evidence recently supplied by Gene Grossman and Alan Krueger and cited by several authors in this volume, the hypothesis that income and environment are positively related cannot be refuted.

A second and somewhat related theme of free market environmentalism is that trade in environmental amenities evolves as consumers demand more of the good. Of course, in order for this to happen, suppliers must have secure property rights to the environment that they can sell to the demanders. If such property rights can be hammered out, environmental resources can be turned into assets that are husbanded. Eco-tourism is a prime example. Tourists are willing to pay huge sums for safaris, but until recently these sums had little impact on the supply of the environment; it only increased the quality of food, transportation, and lodging. By giving natives a claim to the revenues from elephant hunting and photographic safaris, the government of Zimbabwe has all but eliminated poaching. Similarly, Amazonian governments are searching for ways to give residents a stake in conserving the forests where previously they could only profit by exploitation of the resources. And finally, debt-for-nature swaps represent an innovative market response wherein environmental organizations have been able to control environmental assets more directly.

Though there is no guarantee that environmental quality is inevitable with rising incomes, there is sufficient evidence around the world to give us hope. Certainly, corrupt governmental officials may have an incentive to turn their countries into pollution havens or denuded landscapes, but as seen with the decline of communism, such corruption is at least a little easier to discover given modern communications. With significant evidence that markets offer the most efficient way to produce goods and services, there is no reason to shun such prospects for the production of environmental quality. Trade liberalization in the context of free market environmentalism is the friend not the enemy of the environment.

It is within this context that the chapters in this volume were written. They are not all written by authors who would call themselves free market environmentalists, but they all recognize the important ways in which economic growth can be harnessed to the benefit of environmental quality. Indeed, it is hoped that as trade liberalization proceeds, more researchers and policy makers will explore the prospects for including environmental goods in the trade package. It will no longer do for environmentalists in rich countries to force the people in poor countries to subsidize environmental quality, but by opposing free trade agreements, this is exactly what environmentalists are doing. Trade barriers erected in the name of environmental quality surely keep people in the less developed world poorer and doubtfully improve the environment in the long run. If trade liberalization can be expanded to include trade in environmental assets, wealthy environmentalists can put their money where their environment is, getting the results they want while allowing others to move up the economic ladder.

This volume proceeds as follows. The first chapter by Bruce Yandle provides an overview of the NAFTA including its general linkages to the environment. In Chapter 2, Roberto Salinas-León provides a perspective from Mexico where there is overwhelming support for the NAFTA. It is clear that the NAFTA means more to Mexicans than the prospect of access to U.S. markets; it means the possibility of institutionalizing the Salinas government's free-market reforms that have been stimulating the economy since the late 1980s. Steven Globerman, a Canadian economist, provides a framework for integrating environmental assets into traditional economic growth models. In applying his framework, he reviews the literature on the linkages between trade liberalization and the environment, concluding that there is little support for the contention that the NAFTA will result in environmental harm. Economists Peter Emerson and Robert Collinge examine detailed provisions of the NAFTA and also conclude that it is far more likely to improve environmental conditions in North America. Indeed, Peter Emerson has been a voice of sanity in the environmental community, recognizing the potential for harnessing market forces for the betterment of nature. Chapters 5 and 6 focus more narrowly on two environmental areas where trade liberalization can have positive environmental effects. As a policy analyst for Ducks Unlimited Canada, James Patterson argues that competition through trade liberalization may force the governments of both the United States and Canada to change agricultural policies that have subsidized the destruction of wildlife habitat for several decades. Robert Deacon and Paul Murphy use Latin American examples of debt-for-nature swaps to illustrate the difficulty environmentalists have getting their demands expressed in the marketplace. By reducing transaction costs, free trade agreements potentially can encourage capital flows and better environmental stewardship. Finally, the volume ends on a less optimistic political note. Bruce Yandle argues that the same forces that bring together bootleggers and Baptists in a coalition against the sale of liquor on Sunday can coalesce protectionists and environmentalists in opposition to free trade. Just as the bootleggers can use the virtuous goals of Baptists to promote their ends, protectionists can exploit environmentalism. The difference is that, while the former may get the Baptists what they want, the latter is likely to generate protection without environmental benefits. All the contributors hope that this volume helps policy makers avoid such a result.

1

Is Free Trade an Enemy of Environmental Quality?

Bruce Yandle

Introduction

The North American Free Trade Agreement (NAFTA) will unambiguously raise incomes in the region as people in Canada, Mexico and the United States enjoy the fruits of Adam Smith's gains from trade that come when trade barriers are reduced.[1] On February 24–25, 1992, the U.S. International Trade Commission hosted a symposium on the effects of the NAFTA, where the results of twelve econometric studies were reported and discussed. The summary of the symposium stated: "Despite the different approaches taken in these studies, there is a surprising degree of unanimity in their results regarding the aggregate effect of a NAFTA. All three countries are expected to gain from a NAFTA. The greatest impact will be on the Mexican economy" (U.S. International Trade Commission 1992, vi). Noting that there will be adjustment costs, the Senate testimony of Carla Hills, U.S. Trade Representative, regarding expanded trade with Mexico captures the essence of this thinking:

> We have drawn on a variety of economic studies on the impact of freeing trade and investment between the United States and Mexico. All point to a net positive impact on the U.S. Economy. . . . The analysis dealing with the issue shows net job creation in the U.S. . . . Studies also show . . . that some sectors might face increased competitive pressures. . . . Dr. Clopper Almon of the University of Maryland, for example, estimates that under FTA, the United States would gain 88,000 jobs and lose 24,000 over 10 years, for a net gain of 64,000 jobs. (U.S. Congress 1991a, 31)

To the extent that trade among these three countries is a precursor of the momentum that will move south into Central and South America, the potential for further gains from trade is enormous.

While the long term prospects are decidedly bright, there is considerable friction between the wheels of trade and the axle of the status quo. The mass of unidentifiable, unorganized people, especially in Mexico, who will benefit from freer trade have little voice and hardly any political muscle. On the other hand, there are some well identified vocal people who may bear adjustment costs if trade expands. They know about the NAFTA and have lobbyists who make it their business to monitor the agreement. Instead of hearing about the millions whose lives will be improved, we hear of hundreds whose current jobs will be threatened. This view is expressed in classic form in the Senate testimony of William H. Bywater, International President of the International Union of Electrical Workers:

> It is our view that this agreement would be a disaster for IUE members and other American workers. . . . At one time, we had 360,000 members. Today we have 165,000. This dramatic decline has been due, in large part, to the movement of production and jobs out of this country. Multinational corporations have fled to low-wage labor markets such as Mexico. It is the American workers who remain employed in these industries who have much to lose from a free trade pact with Mexico. . . . Our members are outraged that the Bush administration is forging ahead with this agreement, given the current recession and increasing job loss in this country. (U.S. Congress 1991b, 48–9)

If these claims are not enough to cripple the agreement, environmental concerns could provide the death knell. The fear is that industrial growth in Mexico will simply increase already significant pollution in that country. Critics see U.S. industries fleeing to Mexico, attracted by a winning combination of cheaper labor and relaxed environmental rules. Hence, both American workers and the environment will suffer.

Is there a sound basis for these fears? Or have special interest groups exaggerated the magnitude of adjustment costs? Just how long and tortuous is the journey to the free trade promised land? Politics dictate that the longer term gain from trade cannot be secured unless the short term pain is somehow addressed.

This chapter focuses on this adjustment process especially as it relates to the environment. The next section presents a general discussion of free trade. After considering two competing views represented among the critics and supporters of the NAFTA, the third section compares Mexico's past and present trade policies and the emerging NAFTA with the dramatic changes that have occurred in Europe. The fourth section examines the extent to which U.S. industries will be affected by new competition from Mexico. The final section brings together the findings of two important research efforts regarding environmental control and import competition and presents the case for free markets and property rights.

Static and Dynamic Views of Free Trade

The statements by William Bywater and Carla Hills in the introduction pointedly illustrate the static and dynamic views of expanded trade with Mexico. The static view assumes that the world changes very little. Accordingly, increased imports displace domestically produced goods and force specialized workers into unemployment. In its extreme form, the static view sees the current mix of skills, production techniques, industries and firms permanently embossed on the economic landscape. Opening a previously closed trade door upsets the status quo. Lost jobs, bankrupt firms, unemployed workers and declining industries can be quantified to demonstrate the negative impacts of trade.

The dynamic view of expanded trade sees economic activity as a smoothly operating process, not a frozen mosaic. Instead, the economy is continually being reshaped by the actions of purposeful people seeking to better themselves. Competition creatively replaces lower valued, inefficient producers with higher valued, efficient producers. Under this view, there is no particular reason to preserve an industry, firm, or occupation if better opportunities lie in the offing. People enter and exit labor markets; firms expand and contract; new products and production processes emerge; and wages and prices rise and fall. Of course, there are adjustment costs, but these are small when compared to the potential gains.

Though both views recognize the potential for gains from trade, the different policy implications can be illustrated with a simple example. Gains from trade emerge from specialization when the butcher specializes in meat cutting and the logger specializes in wood cutting. The butcher gets wood at a lower cost; the woodcutter enjoys more fresh meat. Because both individuals are made better off, we would expect opposition to governmental policies that disallowed trade between the two and required all wood cutters to produce their own meat and all butchers to cut their own wood. But if woodcutters discover a lower cost meat supply, the static view generates different policy implications. Butchers are no longer likely to sing the praises of free trade when the winds of competition force change.

The dynamic view, however, recognizes that changing technologies and new products will require transition. If a lower cost substitute for the butcher's meat is found, the butcher has several options. He can try to meet the competition, produce a better product himself, or go into some totally different line of work. Though any of these options is costly to the butcher, denying the new competition is costly to all consumers. Protecting the butchers from competition

simply perpetuates higher cost-lower value production and denies overall gains
from trade.

We can see the impact of these competing views of trade in the histories
of the United States and Mexico. Indeed, Adam Smith (1937, 538) noted these
differences in 1776.[2] The American colonies represented a new dynamic trade
order built on expanded freedom, private property, the right to contract, and the
absence of an authoritarian government. Central and South America, on the
other hand, perpetuated the static trade order that had prevailed for centuries in
Europe and, until very recently, prevailed in Mexico. Jobs and industries were
protected. The most vulnerable or politically valuable industries were
nationalized. The few gained much; the many gained little. Competition was
directed from above. Of course there was change and mobility of people, but
countless people followed traditional paths that set narrow limits on their
chances for a better life.

Though the United States has drifted toward more static trade policies, the
Salinas government in Mexico recently has liberalized trade, reduced regulation
of foreign investment, privatized major economic sectors that were previously
nationalized, and laid a foundation for market-directed economic activity.[3]
These policies are in sharp contrast to those set in motion by the OPEC
embargo of 1973 which set a stage for Mexico's political economy of the 1970s
built on the nation's abundant supply of nationalized oil. Following the
embargo, huge amounts of wealth moved through political hands. Governmental
debt increased, inflation rose, and economic growth declined. In an effort to
protect the industrializing sectors from foreign competition and ownership, the
government instituted a regime of tariffs, quotas and restrictions. The Mexico
economy bogged down to a slow process of stagflation.

The Salinas reforms coupled with the NAFTA provide trading potential
that exceeds that of the European Community. Viewed as an open trading
community, the new Europe will have a population of 365 million and a per
capita GNP of $12,931, all based on 1989 data.[4] Compare that to the emerging
open market for Mexico, Canada, and the U.S., which will have 359 million
people and a per capita GNP of $15,545. NAFTA combines "the world's
largest, eighth-largest, and thirteenth-largest economies into a common
economic space" that will rival that of the new Europe (Morici 1991, 45). But
the important question is: Will we take advantage of these dynamic trade
opportunities or revert to static trade policies?

Dropping the Barriers: An Inch, Not A Mile

Much of the debate over the NAFTA contends that opening trade with Mexico
will flood the United States and Canada with cheap Mexican goods, but the evidence suggests that the North American economies are already heavily integrated. For example, Mexico is the third largest U.S. trading partner, accounting

for 6 percent of U.S. imports and 7 percent of U.S. exports.[5] However, given the small relative size of the Mexican economy (3.6 percent of the U.S. GDP in 1989), the United States' share of Mexico's total imports reaches a whopping 67 percent. Given these dimensions, Mexico's expanded trade will have far more impact on the Mexican economy than on the U.S. economy.

What About Current Trade Barriers? As the U.S. International Trade Commission puts it, "with few exceptions, both countries already have relatively low tariffs and nontariff barriers to trade with each other" (U.S. International Trade Commission 1991a, vii). In a real sense, the tariff walls have already fallen. Mexico's move to liberalize trade came in 1986 when the country became a member of GATT. As a condition, tariffs had to be reduced and quotas dropped. In 1989, Mexico's trade-weighted average tariff stood at 10 percent of the value of goods received, down from 25 percent in 1985 (U.S. International Trade Commission 1991a, 1-2). In addition, import licenses previously required for all imports into Mexico now apply to only 230 of 12,000 items. In short, the Mexican trade barriers are low and falling.

How much protection from Mexican goods does U.S. industry currently receive? In 1989, the trade-weighted average of tariffs on imports from Mexico was 3.4 percent (U.S. International Trade Commission 1991a, 2-2).[6] About 9 percent of Mexican imports enter duty free, and another 45 percent originate under the Generalized System of Preferences, which means the value added in Mexico to United States produced inputs enters tariff free.

Of ten major economic sectors potentially impacted by free trade with Mexico, only a few are likely to be affected in a measurable way. Consider them in turn.

Agriculture

Contrary to many arguments, it is agriculture rather than industry that will be the major affected sector. Again, average tariffs are low with the trade-weighted average imposed by the United States equal to 7 percent ad valorem and that imposed by Mexico equal to 11 percent. But both countries impose limitations based on a combination of tariffs and licenses or market orders.

Within agriculture, U.S. horticulture products are the most vulnerable to Mexican competition. The average tariffs are high on both sides of the border: 35 percent for the United States and 20 percent for Mexico. Moreover, the U.S. system of market orders imposes outright limitations along with tariffs. Since Mexico is a lower cost producer of citrus products and winter vegetables that require stoop labor and the United States is a lower cost producer of processed food products, free trade will impose adjustment costs on agriculture in Florida, California and other Sun-belt states. As will be seen for a few other sectors, lower paid, unskilled labor will bear the burden of change for horticulture. Oddly enough, much of that labor in the United States is currently Mexican and migrant labor from elsewhere.

Though horticulture in the United States will contract, other U.S. agricultural sectors will gain from expanded trade. For example, producers of grain and oilseed should see an expansion of shipments to Mexico, as would producers of meats.

In summary, the rub in U.S. agriculture hits primarily U.S. producers of fresh vegetables. Opportunities for lower paid workers will be reduced; higher paying jobs in canning and processing will expand. Researchers for the U.S. International Trade Commission examined the overall effects of open trade on the U.S. agricultural sector and found that "The top job-gaining sectors in the United States are agriculture, machinery and metal products. Total U.S. employment increases by 44,500 jobs after five years, with the largest gains (10,600 jobs) in agriculture" (U.S. International Trade Commission 1992, 15–16).

Steel

The U.S. steel industry is often cited as a potential victim of open trade with Mexico. Already 15 percent of the steel produced in Mexico is imported into the United States accounting for about 14 percent of all U.S. imports in 1989.[7] However, Mexico's shipments have been limited since 1985 by a system of voluntary restraint agreements to 1 percent of the U.S. market, and the quota has seldom been binding.[8] Even if Mexico shipped all its steel to the United States, that would amount to only 12 percent of U.S. consumption since the U.S. internal market is so large. On the other hand, the United States is a major supplier of steel to Mexico, with exports amounting to 1 percent of total U.S. production in 1989.

These aggregate data disguise important features of the U.S. and Mexican steel economies. Steel marketing and ownership are global. For example, Mexico's only stainless steel producer, Mexinox, is owned partly by Thyssen Edelstahl of Germany and Acerinox of Spain.[9] North American buyers receive 60 percent of the firm's output. In the United States, there are 63 production facilities with extensive foreign ownership (U.S. International Trade Commission 1991b, 3-10, 3-11). The countries involved include Japan, France, South Korea, Sweden, the United Kingdom, Italy, Brazil, Germany, Canada, China, Switzerland, and the Peoples Republic of China. When one speaks of the U.S. steel industry, one is speaking of truly multi-national ownership.[10]

If employment effects in the steel industry are the crux of the issue, the outlook under the NAFTA is decidedly favorable. Open borders will lead to greater U.S. exports to Mexico and expanded shipments to the United States from Mexico, but on balance, the United States will gain more market than it loses (U.S. International Trade Commission 1991a, 4-37).

Chemical

The U.S. chemical industry is another sector that some suggest will experience dislocations if trade is opened with Mexico. In this case, the dislocations are peculiar to the U.S. petrochemical industry, the world's largest supplier. At present Mexican law prohibits foreign investment in petroleum related industries. Simply, the United States is locked out of direct investments in Mexico. U.S. law prohibits the exportation of petroleum, and Mexican law limits the importation of natural gas. Recent modifications in the investment restrictions opened the door a bit, but meaningful limitations remain in effect.

The overall impact of the NAFTA on the U.S. chemical industry is likely to be trivial, other than for investments that might be made in production facilities in Mexico. Such facilities could compete with U.S. producers, but this is unlikely to be a major problem given that the U.S. chemical industry exports far more to Mexico than is imported. Tariffs are low at approximately 4 percent. An open border would not likely disturb U.S. employment patterns or capacity utilization.

An exemption is the U.S. pharmaceutical industry which would gain substantially if the NAFTA led to solid enforcement of intellectual property rights. U.S. patents would then be protected, and the U.S. industry would expand shipments and production across the border.

Other Industries

A review of other import sensitive industries generates little evidence of significant NAFTA impact. Textiles, apparel, electronic products, metal working machinery, and machine tools all present different cases, but there is little to suggest major dislocations. In the case of textiles, the effects are almost neutral. For machine tools and metal working machinery, however, the effects are decidedly favorable.

Producers of lower priced, common household glassware are predicted to suffer significantly from open trade. That industry has been buffeted steadily by international competition and is still in a state of transition. It is currently protected by a 22 percent tariff and faces a 20 percent tariff for exports to Mexico. If the tariff walls go down given the size of the U.S. market relative to that of Mexico, shipments to the U.S. will increase markedly.

The best estimates of the impact of the NAFTA indicate that, on balance, the effects will be positive. Overall employment will expand. Incomes will rise. Producers of fresh vegetables, citrus products, glassware, and some steel products will feel the cost of adjustment. When it gets down to particular farms, plants and towns, the NAFTA will generate change. In some cases, change has been waiting in the wings for a long time, and tariffs and quotas have simply delayed the inevitable. In their absence, adjustments would have come smoothly, continuously, not all at once. When the pink slips come, displaced workers will understandably point the finger of blame at the NAFTA. Special interest groups whose incomes depend on controversy will make certain the NAFTA is not forgotten.

The Environmental Linkage

There are two ways in which the NAFTA is linked to environmental quality. The first is through its impact on relocation decisions, and the second is through its impact on the demand for environmental quality. Fortunately, significant empirical research is available for both.

In its analysis of the connection between free trade and the environment, the Office of the U.S. Trade Representative developed a "hit list" of industries vulnerable to the intertwined forces of reduced tariffs and high cost pollution control (Office of the U.S. Trade Representative 1991). To be on the list, an industry had to face high environmental compliance costs, be exposed to expanded import competition from Mexico, have low relocation costs relative to U.S. environmental compliance costs, and find lower environmental costs in the new location.

After examining 445 U.S. industries, the analysts found 11 industries vulnerable to the effects of environmental rules, reduced tariffs, and relaxed investment restrictions. The "hit list" industries are specialty steel, petroleum refining, five categories of chemicals including medicinal compounds, iron foundries, blast furnaces and steel mills, explosives, and mineral wool (Office of the U.S. Trade Representative 1991, 141). Probing deeper, the Commission's report notes that 10 of the 11 industries have high capital intensity, thus reducing the likelihood that plants will relocate to take advantage of lower environmental costs in Mexico. The gray and ductile foundry industry is the single industry that best meets the four-part criteria. The report emphasizes that there is no case for wholesale outmigration of U.S. plants to Mexico. The facts simply do not support the argument.

But there may be other costs that enter the relocation matrix. For example, the report also considered the cost of liability rules (Office of the U.S. Trade Representative 1991, 142). Since the Bophal disaster brought hard times to Union Carbide even though the firm was meeting India's standards, multi-national firms have taken a very conservative environmental stance. Rules of law cause them to base their foreign environmental standards on U.S. rules. Indeed, Ford Motor Company's written policy for plants in Mexico exactly incorporates U.S. environmental standards (Office of the U.S. Trade Representative 1991, 142). The important point to note is that something other than government regulation can discipline firms in their use of the environment (see Meiners and Yandle 1992). Common law courts can and do impose liability standards that may affect environmental quality more than bureaucratic regulations.

Finally environmental quality may improve because new plants tend to use the latest technology and equipment which reduce inefficiencies and pollution. Even if the dirtiest U.S. plants that cannot afford to meet U.S. standards move south, they will operate new plants.

Next we turn to the question of what determines the demand for environmental quality. What causes environmental rules, property rights enforcement, and careful natural resources management to emerge? Can we find a reliable driving force that logically leads to improvements? Or must we rely on the fragile forces of ethics, religion, and environmental awareness?

The heated debate over the impact of free trade on the environment led Princeton economists Gene Grossman and Alan Krueger (1991) to measure the relationship between environmental quality and higher incomes (industrial growth). Under the sponsorship of the prestigious National Bureau of Economic Research, they examined the relationship between urban air quality and economic growth for 42 countries. Specifically, they focused on sulfur dioxide and suspended particulates. Of course, there is more at stake in Mexico and the United States than air quality, but this provides a reasonable proxy for environmental quality in general. Their sample included countries with annual incomes per capita ranging as low as $1,000 and as high as $17,000. Their statistical models also adjusted for government form (communist-noncommunist), population density, and other characteristics that logically affect air pollution.

The Grossman–Krueger results indicate that air pollution rises with income when incomes are at the very lowest levels of $1,000 to $5,000. Once incomes are above $5,000, however, air pollution falls, and it continues to do so as incomes rise. As an aside, they found that communist countries have systematically higher levels of pollution, all else equal.

The authors point out that the implications for Mexico are rather straight forward. Mexico is approaching the threshold. Consistent with the Grossman–Krueger results, Mexico has begun to increase its environmental standards. Since higher levels of income yield even more improvements, the NAFTA should generate greater demand for environmental quality in Mexico.

Final Thoughts

Is the NAFTA an enemy of environmental quality? The evidence presented here emphatically suggests that it is not. Of course, the story is complex, and the outcomes of expanded trade with Mexico will surely generate instances of environmental degradation. But instead of viewing trade through static glasses and isolated cases, we must focus on the broad general tendencies that emanate from the dynamic economy. We must consider the case without expanded trade. What would be the state of environmental quality if trade were reduced? There is strong evidence that it would deteriorate.

We must recognize that free trade itself guarantees very little. What is necessary is a system of well established property rights to natural resources and environment. If individuals have the security of property rights protected by a rule of law, each individual has no choice but to account for the costs of his actions. Damage to property, be it environmental or otherwise, is a cause of action against the damaging party. Freer trade under the NAFTA will raise incomes and with them the demand for environmental quality. The challenge for all three parties to the agreement is to provide the stable set of property rights to the environment that will encourage good stewardship. Unfortunately, there is good reason to expect that special interests will mitigate against this outcome and press instead for command-and-control environmental regulations that will erode potential gains from trade possible with the NAFTA (see Chapter 7).

Notes

1. Also in keeping with Smith, there are supporters of the NAFTA who, in pursuit of their own self interests, applaud the prospects of expanded trade in Mexico. For a sampling of these, see Green (1991, 3A); *Paperboard Packaging* (1992, 10); *Daily News Record* (1992, 11); and Denver (1992, 10A).
2. For a more complete discussion of the new and old orders, see Novak (1982).
3. It should be noted that from World War II until the 1970s, Mexico's economy grew annually at the rate of 6.7 percent, far exceeding its population growth and the GNP growth of its neighbors. For more on this, see Morici (1991). Also see, U.S. International Trade Commission (1991a, 1-1, 1-2).
4. The data here are developed from Morici (1991, 45).
5. The data in this section come from U.S. International Trade Commission (1991a, vii).
6. According to a study of the Office of the U.S. Trade Representative (1991, 138–9), tariffs do reach 22 percent for copper and 10 percent for many steel products.
7. This section draws on data in U.S. International Trade Commission (1991a, 4-35 through 4-37).
8. The VRAs came in negotiations after Mexican steel producers were charged with dumping under U.S. trade law. To avoid penalties, Mexican producers agreed to limit shipments. In a real sense, a cartel is formed that raises the price of steel for U.S. consumers and producers. A complete free trade agreement would have to chisel away VRAs.
9. For discussion, see U.S. International Trade Commission (1991b).
10. The picture becomes even more confusing when actions are taken by "U.S." firms to stop foreign producers from selling steel in the U.S. at prices below those charged in the home market. Recently, several major U.S. firms filed a petition with the ITC and the U.S. Department of Commerce alleging such dumping. According to press reports, the charges are filed against "all trading partners," including Canada and Mexico. Of the U.S. firms filing the complaint, three have large foreign ownership participation. One is totally owned by a Japanese steel maker, which will be charged with dumping (*Wall Street Journal* 1992, A6).

References

Daily News Record. 1992. Say stores value Mexican sourcing. (January 16): 11.

Denver, John. 1992. Trade promoter shifts focus from Far East to South of the Border. *Journal of Commerce* (February 11): 10A.

Green, Paula L. 1991. U.S. liquor exporters pin hopes on trade pact. *Journal of Commerce* (December 5): 3A.

Grossman, Gene M., and Alan B. Krueger. 1991. Environmental impacts of a North American Free Trade Agreement, working paper no. 3914. Cambridge, MA: National Bureau of Economic Research, November.

Meiners, Roger E., and Bruce Yandle. 1992. Constitutional choice and the control of water pollution. *Constitutional Political Economy* 3(3): 359–80.

Morici, Peter. 1991. *Trade talks with Mexico: A time for realism.* Washington, DC: National Planning Association.

Novak, Michael. 1982. *The spirit of democratic capitalism.* New York: Simon and Schuster.

Office of the U.S. Trade Representative, Interagency Task Force. 1991. *Review of U.S.–Mexico environmental issues,* draft. Washington, DC, October.

Paperboard Packaging. 1992. Free trade with Mexico to benefit suppliers. (January): 10.

Smith, Adam. 1937 [1776]. *An inquiry into the nature and causes of the wealth of nations.* New York: The Modern Library.

U.S. Congress. Senate. 1991a. Statement of U.S. Trade Representative, Carla A. Hills. Hearings before the Committee on Finance, May 7.

U.S. Congress. Senate. Committee on Foreign Relations. 1991b. *Issues relating to a bilateral free trade agreement with Mexico.* Statement of William H. Bywater, International President, International Union of Electrical Workers. Hearings before the Subcommittee on Western Hemisphere and Peace Corps Affairs, March 14, 22 and April 11.

U.S. International Trade Commission. 1991a. *The likely impact on the United States of a free trade agreement with Mexico,* USITC Publication 2353. Washington, DC: U.S. International Trade Commission, February.

———. 1991b. *Steel industry annual report,* USITC Publication 2436. Washington, DC: U.S. International Trade Commission, September.

———. 1992. *Economy-wide modeling of the economic implications of a FTA with Mexico and a NAFTA with Canada and Mexico,* USITC Publication 2516. Washington, DC: U.S. International Trade Commission, May.

Wall Street Journal. 1992. U.S. steelmakers launch suit against foreign firms. (June 29): A6.

2

Free Trade and Free Markets:
A Mexican Perspective on the NAFTA

Roberto Salinas-León

Introduction

The recently signed North American Free Trade Agreement (NAFTA) has been the subject of a wide array of public policy debates in the United States and Canada. Notwithstanding the immense socioeconomic benefits of free trade, these debates are unsurprising given the commercial interests especially in the United States and politics involved.[1] What is surprising, however, is the interesting new protectionist coalition that has emerged in a more sophisticated guise.

Nevertheless, the discussions about the pros and cons of the NAFTA embody highly specific (and highly peculiar) characteristics. One salient irony is that the overwhelming majority of debates and discussions have centered on strictly extra-commercial concerns: job losses due to trade liberalization; environmental degradation resulting from greater trade flows with Mexico; democracy and human rights; and even issues of cultural identity and religion. Absent from these arguments are several important questions. Is the NAFTA trade diverting or trade creating? Are the established phase-out periods sufficient to avoid industrial and commercial disadjustment? Are the rules of origin, dispute settlement and intellectual property rights acceptable for the purposes of expanded trade?[2]

Another equally salient irony about the current politics of the NAFTA is that it has gained far broader support in Mexico, a country with a long statist and protectionist tradition, than in the United States and Canada, supposed bastions of free trade and free markets. Indeed, the NAFTA has become a centerpiece of Mexico's everyday culture. Every newspaper and magazine features a story about the accord. Most citizens are relatively well-versed on its implications. And polls consistently reveal that 70 to 80 percent of the country's citizens respond positively to the challenges of free trade with their mighty northern neighbor.[3] In effect, the NAFTA has become symbolic of a profound positive transformation in Mexican society.[4]

In this paper, I will succumb to the arguably irresistible temptation to focus on the extra-commercial implications of the NAFTA and instead aim to answer two fundamental questions:

1. Why is the NAFTA so well-received, in a society with an established reputation for protectionism and state interventionism?
2. Why does the NAFTA represent a promise for progress and prosperity in Mexico?[5]

My argument is that the NAFTA plays a crucial strategic role in Mexico's current public policy because it seals the permanence of responsible government and market-oriented change. This suggestion is important for assessing the impact of the NAFTA on trilateral grounds, lest the standard arguments espoused north of the Rio Grande remain unduly centralized and incomplete. I shall conclude with reference to the widespread but ill-founded concerns surrounding the environmental effects of the NAFTA in Mexico.

Nationalism and New Nationalism in Mexico

In his third presidential address in November of 1991, President Carlos Salinas de Gortari expounded on the shift from former nationalism to what he labeled "the new nationalism." Under his definition, new nationalism is that which is good for the nation. For example, if privatization and free trade are good for the nation, they qualify as "nationalistic." This new nationalism, so construed, has been an important instrument to short-circuit criticisms from the left that the Salinas government is selling off the family jewels to the gringo rivals of old.[6]

While the new nationalism may seem redundant, it marks a dramatic turnaround in the standard, paternalistic interpretations of "national sovereignty." It also harbors an important clue as to why the NAFTA enjoys such wide popular support in Mexico. The "old nationalism" is reminiscent of the crisis-ridden years of the 1980s, during which the country's currency experienced a 47 percent loss in real purchasing power and a 23,000 percent devaluation. Indeed, much economic havoc has been perpetrated in the gallant name of "national sovereignty." This has both an historical and contemporary basis.

Mexico enjoyed three decades of sustained and stable economic growth prior to the advent of import-substitution policies and massive state intervention in daily economic life. In 1970, the country averaged 6 percent growth rates, low inflation rates of 4 percent, and a meager foreign debt of $4 billion. The real exchange-rate stood at 12.50 pesos per $1 (U.S.) dollar. In 1970, however, the Echeverria Administration reversed the strategy of "stabilizing development" and began a program of import-substitution and massive state intervention. This was justified on the grounds of the need to attain "national sovereignty" and a more "equitable" distribution of wealth.

The ensuing years witnessed a dramatic rise in state-run enterprises (going from 300 in 1970 to 1,200 in 1982) and an explosion in the bureaucratic apparatus of some 400 percent. In 1982, the Lopez Portillo Administration announced default on external debt interest obligations, drastically devalued the currency by 90 percent, and expropriated all banking assets. By the end of 1987, inflation skyrocketed to 159 percent, foreign debt escalated to $120 billion, the second largest in Latin America, and the exchange rate depreciated to 2,800 pesos per $1 (U.S.) dollar. Negative growth rates hence became the rule rather than the exception. Not surprisingly, capital flight increased exponentially, with an estimated $80 billion finding an investment home overseas during the 1980s. Because free trade provides the antithesis of the statist and protectionist policies that produced the "lost decade" of 1980s, Mexican citizens and the Salinas government have embraced the NAFTA as the only viable alternative to overcome rampant poverty and stagnation.

The Makings of a Modern Mexico

Mexico has experienced dramatic changes in economic and structural reform since Salinas de Gortari assumed office in 1988. It has repudiated excessive state intervention and encroaching protectionism, embracing open markets and open commerce. Formerly a perpetual land of "mañana" with bleak hope for a brighter future, Mexico is now forging a dynamic policy of globalization, competitiveness, and sustained economic growth. Last year, despite an unfavorable world economy, the rate of growth registered 2.7 percent, and surpassed the rate of demographic growth for the fifth straight year.

Mexico's newfound prominence in global change reflects a positive reputation as a country with a superior investments regime. In 1992, it experienced an inflow of over $10 billion in foreign investments—thereby augmenting hard currency reserves to a high of $18 billion. The once-fantastical target of $24 billion for the Salinas presidential term has already been met, two years ahead of schedule. In 1988, total foreign investment was 5 percent of GDP. Today, this figure has grown to 15 percent of GDP.

This autonomous improvement in confidence and the climate of investment, reflects the success of revolutionary transformations designed to give the private sector a pivotal role in revitalizing and securing growth. To this end, the government has privatized and liberalized several areas of the economy previously considered the exclusive domain of state management, including telephones, highways, the banks, mining, agriculture, pensions, water, and housing. This process of internal reform has been complemented with aggressive trade liberalization, which has made domestic businesses more competitive and more diverse. Ten years ago Mexico's principal exports were drugs, illegal workers, and populist rhetoric. Today, they are refrigerators, aluminum, computer keyboards, electronic parts, and auto engines.

In effect, free markets and free trade have become the nation's new economic way of life, and the pace and process of reform have been significant. The program of structural change has developed as a four-dimensional package, based on fiscal discipline, deregulation, privatization, and multilateral commercial opening. It is an integrated program which has successfully put forth a frontal attack on the economic disequilibrium inherited from the 1980s.

In the rubric of fiscal discipline, the government boasts several landmark accomplishments, including balanced budgets and fiscal health. In 1992, it registered a budget surplus of 0.4 percent of GDP in federal finances, its first in modern times. It has a stronger projection for 1993, of 1.7 percent of GDP, which does not include windfall revenue obtained via the sale of remaining state enterprises. The fiscal deficit was 16.9 percent of GDP only four years ago. The consistent practice of fiscal discipline via austerity and substantial debt amortization (foreign and domestic) have enabled the present government to heal public finances, fortify investment flow expectations, and halt runaway inflation. Indeed, the inflation rate has been drastically reduced, from 159 percent in 1987 to 11.5 percent last year. Now, single-digit inflation rates, something the country has not enjoyed since 1971, will become reality.

In the rubric of deregulation, critical initiatives have been taken in sectors including transport, highway construction, aquaculture, and private provision of potable water services.[7] In 1992, the government authorized a private pension program for worker retirement funds, similar in spirit to the highly successful liberalization of social security services in Chile. This will improve the quality of retirement funds for Mexico's vast labor force, and serve as a catalyst for strengthening domestic savings. Most recently, a broad-based initiative to deregulate housing was introduced, which includes provisions to permit full foreign ownership and investment of housing services.

Mexico enjoys one of the world's most important privatization programs which has successfully privatized 926 state-run companies. Recent landmark privatizations include the telephone company, steel mills, ports, banks, and fertilizer and mining concerns. Overall, the privatization program has had a three-fold impact on the efforts to revamp the economy. First, it has divested unprofitable concerns and thereby helped to heal public finances. A case in point is the sale of the steel companies, AHMSA and SICARTSA, which formerly absorbed up to $700 million annually to finance their inefficient activities. Second, it has improved business performance. A salient example of renewed efficiency is the airline, AREOMEXICO (formerly AEROMAYBE due to terrible on-time departure rates), which in a small space of three years went from one of the world's worst to one of the world's best. It has reached record performance in on-time departure and in baggage handling. Third, many privatization efforts have contributed toward spectacular stock market growth (currently the eighth most attractive worldwide, notwithstanding highly volatile behavior in 1992). The international placement of TELMEX (L-type, non-voting) shares, for example, attracted unprecedented sums of capital investment.

The modifications of constitutional provisions in Article 27 to allow property ownership of ejido land plots is arguably the most significant undertaking in privatization efforts to date. The new legislation is crafted to stimulate several joint ventures between business and ejidatarios, all across the country. They are called "associations of participation" and are inspired by the enormously successful joint venture in the region of Vaquerias, in Nuevo León.[8]

In 1993, the government has announced an ambitious privatization program of port and airport services, provision of potable water, infrastructure projects, and companies such as the insurance conglomerate, ASEMEX, the largest of its kind in Latin America. Overall, the privatization initiative has reduced the number of state entities from 1,200 in 1982 to 260 ten years later.[9]

The fourth item of economic reform, trade liberalization, has been vital to the economic policy that has injected much dynamism in Mexico's global trade performance. A crucial (and often unacknowledged) trait of the country's new trade regime is its multilateral character: open borders north and south, east and west. In addition to the NAFTA, a free trade agreement has already been signed with Chile, and many similar agreements are underway with numerous Latin American countries. This, in turn, has set a healthy precedent in the region. In effect Mexico is leading its southern neighbors in the realization of the Enterprise for the Americas, which foresees the creation of a vast free-trade region spanning from Alaska to Argentina.[10] One of the most important aspects of Mexico's multilateral trade policy is that it avoids the dangers of limiting Mexico's participation in one regional trade bloc, free on the inside but protected on the outside. Most certainly, the signing of the NAFTA signals the end of decades of bilateral uneasiness between the three North American countries.

The Mexican Economy Under the NAFTA

Mexico's uncontested progress in economic reform and liberalization of trade, has made the goals of reaching a free economy a fully realizable option. Yet, while much has been done, much also remains to be done. The inflation rate still needs to come down to guarantee monetary stability. In addition, a stronger effort to de-bureaucratize economic activity and simplify tax structures is required to harness the economic energies required to compete in an increasingly sophisticated and demanding international trade market.

In fiscal and investment matters, the challenge remains to ensure stability and confidence. This requires provisions to allow more foreign flows of investment, already needed to finance a growing current account deficit, as well as fiscal incentives to guarantee that incoming capital stays home. In addition, labor law reform is required to improve the institutional conditions for successful small and medium-size business competitiveness. Today, a small business must spend an amount equivalent to capitalization costs to meet a burdensome 1,200 labor regulations.

Perhaps the greatest institutional challenge lies in the arena of private property ownership rights. Such rights are the institutional basis of a free and thriving economy, yet are sorely absent in the nation's legal framework. In particular, structural reform in the "economic chapter" of the Constitution (Articles 25–28) is required to eliminate uncertainty and provide full guarantee of ownership rights. A stable and sustainable rate of development is unlikely to ensue under legal provisions of state rectorship and clauses which allow government direction of economic affairs in all sectors of society.[11]

The NAFTA will help sustain the economic progress that has been made in the last few years in four fundamental ways. First, it will reduce transaction costs by providing a more formal legal structure of the flourishing trade that already exists. This legal environment will include mechanisms for dispute settlement. Provisions governing intellectual property or trademark rights are included in the NAFTA and were necessary to by-pass problems of piracy and to open up each nation's substantial service markets. Other investment-related provisions reduce transaction costs by ensuring full regional access to capital flows.

The second argument for the NAFTA in Mexico is diversification. In effect, open trade liberalization has diversified Mexico's export sector away from dependence on oil. Manufactured goods are growing at a steady rate of 15 percent per annum, and represent 55 percent of the country's total external output. In 1991, despite the fall in international oil prices, Mexico consolidated its export potential by becoming the twentieth most important export nation worldwide. In contrast, prior to trade liberalization, oil exports dominated the external sector by some 75 percent. Today, oil sales abroad constitute less than 30 percent of all exports. So seen, the NAFTA constitutes an opportunity to further diversify Mexico's exports and to attain higher levels of domestic competitiveness via duty-free market access.

The concept of competitiveness supplies the third fundamental argument for the NAFTA. Mexico's private sector has begun to adapt to global economic change and the challenges of world integration. The NAFTA is crucial to forge a strong and sufficiently competitive business sector, able to compete worldwide, penetrate new markets abroad, and maximize comparative advantages. The NAFTA lays the foundations of competitiveness, by providing broad-based rules to allow for long-term planning, access to updated technology, incentives to specialize, and free entry to the largest market in the world.

The trend toward modernization is already visible. With trade liberalization, Mexico has recorded explosive current account deficits, as a result of an increase in private investment inflows of $21 billion in 1992, 6 percent of GDP. The high surpluses in the capital account have been used to finance intermediary durables and capital-intensive goods, which together account for some 88 percent of total net imports. This is the same pattern which other nations such as Germany, Spain, Portugal, Japan and the Pacific Rim "tigers" have followed.[12]

The NAFTA represents an opportunity for Mexico to increase its competitiveness in world markets in other ways. Though the existing Generalized System of Preferences enables many Mexican goods to enjoy duty-free status in the U.S. market, trade quotas and other quantitative and technical restrictions inhibit a number of Mexican firms from reducing costs by capturing economies of scale. The NAFTA stipulates that 84 percent of all Mexican goods (some 7,300 products) will receive full duty-free and quantitative-free treatment, as of January 1, 1994. By phasing out duties and other restrictions on 40 percent of U.S. and Canadian goods, the NAFTA will increase Mexico's competitiveness by making high technology and modern equipment more available. In this way, Mexico will have access to capital and intermediate products free of tariff restrictions and thereby accelerate the urgent process of modernizing the productive plant. Medium and small-size companies will gain an opportunity to specialize at far greater degrees.

The possibility of improved opportunities from being more competitive also mean that many firms will have to become lean and mean. Firms which grew and maintained pace under high-tariff protectionism must now redeploy their assets and modernize their operations. Otherwise, the normal demands of an open consumer market will lead them to bankruptcy. This is a constant source of fear among Mexico's business sector, but it is a fear they accept.

To be sure, competition will stiffen significantly, but the number of new opportunities defies comment. After all, a trade arrangement like the NAFTA expands the consumer market for Mexican firms from a relatively impoverished 80 million consumers, to a relatively rich 360 million. Of course, to reap the full benefits of unrestricted trade and meet the several challenges of regional competition, Mexico's internal private sector will require large investment in human capital and development. A modern approach focusing on training and educational skills is essential to cultivate a competent labor force capable of meeting the requirements of competition and trade performance.

A fourth and potent argument for the NAFTA in Mexico is that this accord will balance out the capital needs of a severely undercapitalized economy. The financial burdens of reconstructing the economies of Russia and Eastern Europe, the high cost of unifying the two Germanys, and the ferocious fight for new capital flow in underdeveloped nations are factors that translate into a fiercely competitive scenario for attracting global investment.[13] If only 25 percent of the world capital stock is destined to the less developed world, it is imperative for Mexico to continue attracting the capital necessary to finance investments estimated to be $150 billion per year for the next decade. In a sense, the NAFTA arrangement was negotiated to meet this challenge. It is less a trade accord than an investment strategy intended to generate resources to finance new jobs. This is an extra-commercial aspect of the accord, albeit one that is crucially important to the economic strategy in place.

With the NAFTA, Mexico stands on the threshold of becoming a major force in global trade. This would enable Mexico to enjoy membership in the largest market in the world, with $6.2 trillion in economic product and a consumer market of 360 million people. This market is 25 percent larger than the twelve nations that make up the EC. The United States already is Mexico's most important commercial partner. In 1992, two-way trade exceeded $70 billion, almost triple the amount in 1986 when Mexico jointed GATT. Seventy percent of Mexico's imports come from the United States, and some 65 percent of its exports find a home in U.S. markets. Similarly, 64 percent of total foreign investment comes from the United States. Mexico represents a very attractive market for U.S. exports and is already the United States' third largest trading partner, ahead of powerhouses like the United Kingdom and Germany. In 1992, Mexico overtook Japan's place in manufactured trading, and is on the brink of becoming number two in the United States' trading priority. Such facts reveal that a future agreement on trade matters was inevitable.

As Salinas de Gortari is fond of stating, Mexico seeks to "export goods, not people." The NAFTA is designed to provide Mexico with the wherewithal to generate annual growth rates of 6 percent. So construed, the treaty has explosive job creating potential, something Mexico sorely requires to service the 1.5 million workers that join the labor force every year.

But the NAFTA means more than expanded trade opportunities; it is symbolic of the profound transformations undertaken during the Salinas government. It is a supra-national and institutional guarantee that the future of Mexico has changed for the better. The NAFTA, therefore, is both a culmination of the process of trade liberalization and, more importantly, a public policy item which consolidates the process of internal, economic structural reform. In other words, it symbolizes the irreversibility of market-based policies, making changes independent of the political whimsy of future administrations.

Mexico is eager to become part of the First World. In the past, illegal immigration, corruption, paternalism and state giantism hurt the country's potential and prospects for economic growth. Now, the "lost decade is over" and a new epoch lies ahead. Mexico is a nation of 82 million persons, 60 percent of whom are under age 30. To realize its full potential as a leading trade player, however, more work remains to be done, both in entrepreneurial and institutional matters. It is in this latter area where the NAFTA is a fundamental strategic instrument: it forces the government to follow a competitive public policy and to generate the institutional conditions for developing a prosperous society.

The Strategic Side of the NAFTA

Some opponents of the NAFTA construe the accord as culmination of a "fortress" strategy intended to shield the North American bloc from outside competition. This claim is highly dubious, but deserves careful scrutiny. Moreover, by putting it to task, we gain a better appreciation of the tremendous strategic value of the NAFTA for Mexico.

The contention that the NAFTA is a "trade-diverting" attempt to build a regional fortress bloc overlooks the fact that capital investment is a highly scarce commodity. To repeat, the NAFTA was crafted in Mexico to give the country an important advantage over other underdeveloped nations in generating an attractive investment climate. This is seen in Mexico as an all-important feature of the NAFTA. Indeed, in a dramatic expression of the commitment to the treaty and in rhetorical response to claims that the NAFTA should be renegotiated, high authorities have recurrently reassured the public that the text will under no circumstances undergo any changes. In the words of Trade Minister Jaime Serra, "not a comma will change."

The problem remains, nonetheless, whether the NAFTA constitutes an "exclusionary" trading bloc, designed not to expand world trade flows but to restrict them to a single specific region. There are three well-founded reasons why this is not true. One is that all three nations are members of GATT, and GATT allows a trading agreement in large unified zones only if those zones are consistent with open multilateral trade. This rules out discrimination. Another reason is that the NAFTA's text contains an accession clause, potentially allowing for a greater number of nations to join the trading arrangement. So far, Chile and New Zealand have expressed interest. Finally, in the case of Mexico, there is a highly systematic effort to pursue a multilateral trade policy that opens borders north, south, east, and west. In effect, the Salinas administration has expressed the willingness to negotiate, bilaterally or otherwise, with whichever nation is interested.

Note, however, that whatever the virtues of a multilateral trade policy and whatever the investment benefits created by a commercial accord like the NAFTA, the establishment of a continental free-trade region, necessarily requires a proper institutional scheme. It is essential for the debates forthcoming in the U.S. Congress to factor this strategic element of the NAFTA into the analyses and arguments, as legislators now prepare to probe whether such an initiative is beneficial to the United States. Yet, there is no coincidence that trade liberalization has accompanied the best moments in the bilateral relationship between the United States and Mexico. Free trade is the basis of peace.

Regional agreements like the NAFTA embody an often unnoticed strategic benefit in making countries more competitive in economic and monetary policy. Hence, crucial changes in agrarian law, port and airport privatization, private management of highway infrastructure, water deregulation, and much more, constitute results of the salutary effects that trade liberalization has on making extant economic structures more competitive.

Indeed, Mexico's trade liberalization program and great leap forward in multilateral free trade underscore its ambitious reform process. Many structural steps remain in the country's economic regime, but the temporal logic seems to point positively towards the direction of reform in other sectors of society, notably oil and electricity.[14] The strategic value of the NAFTA derives directly from its potential impact in literally forcing government, a characteristically whimsical and arbitrary institution in Mexico, to behave in a globally proper and responsible fashion. This means that a policy which inhibits competitiveness will either disappear or adapt to change.[15]

The consensus among Mexico's populace is that the NAFTA constitutes a device to push the country away from its statist and protectionist past and towards more institutional transformations in society. It is believed, in turn, that this restructuring will make possible the abandonment of Third World status and consequently the amelioration of a poverty-stricken society. This is the NAFTA's fundamental contribution to Mexico's current drive toward modernization.

Curiously, therefore, market access constitutes a secondary, albeit important, concern for Mexico's efforts to pursue closer trade ties under the NAFTA.[16] The treaty holds a promise of progress and prosperity, precisely because it functions as a key device to lock in market-oriented policies and thereby forge a confident climate of investment. In essence, this is the main argument for the NAFTA in Mexico.

Conclusion

The strategic side of the NAFTA and its positive effects in forcing Mexico's government to follow a responsible economic course are virtually absent in most discussions in the United States and Canada relating to the effects of the NAFTA. Yet a broad analysis of the role of the NAFTA in Mexico's economic future reveals that there is much more at stake than merely an option to capitalize on its large pool of "inexpensive labor."[17]

Consider, for example, the sundry environmental concerns surrounding current policy debates on the NAFTA in the United States. Notwithstanding the inclusion of explicit environmental clauses in the text of the treaty (a feature which has won the NAFTA the popular label as "the greenest treaty ever negotiated"), radical environmental groups like Friends of the Earth, Public Citizen, and the Sierra Club continue to decry free trade and the NAFTA as a source of immense ecological ills. The standard arguments is that a NAFTA will increase pollution along the Rio Grande border, as well as drive many U.S. businesses south of the border to take advantage of Mexico's allegedly lax environmental regulations and enforcement practices. In addition, there is much worry that U.S. standards will be compromised for the purposes of greater trade.

It is very difficult to prove that such allegations are, by and large, smoke screens for vested interests (though there is no doubt that opposition to the NAFTA "helped" the AFL-CIO effect an environmental turn). Nonetheless, it is very difficult to resist this conclusion. Whatever the motives, notice that none of the concerns incorporates the Mexican viewpoint. For this reason, such allegations are incomplete.

In addition, they are ill-grounded. Common sense logic suggests that greater trade brings greater economic growth which, in turn, generates greater resources to finance ecological protection programs. And specialized studies bear this logic out with empirical evidence to show that beyond a certain per capita income, a greater environmental attention to pollution and other factors ensues.[18] Similarly, research reveals that lax environmental enforcement generally constitutes a tiny incentive for business relocation. If such were the case, moreover, U.S. companies would be quickly relocating in other countries with even more lax ecological standards.

Mexico has expressly stated that it does not wish to become a pollution haven. To this end, a significant number of steps have been taken to ameliorate the country's admittedly severe environmental problems. These include new legislation modeled upon U.S. standards, as well as a tenfold increase in resources for enforcement. The newly formed Border Environmental Plan, for instance, assigns $180 million annually for the next three years to finance projects of water and waste improvement. The amount is equivalent to roughly 0.5 percent of the country's GDP, far superior in real terms to the resources committed by the U.S. government.

In this chapter, I have argued that the NAFTA is a crucial item for Mexico's attempt to modernize and develop. Yet open trade is not a zero-sum game. Indeed, all arguments for the NAFTA are ultimately reflections of a simple (but universal) truism that the voluntary exchange of goods and services enlarges the size of the pie so that everyone, including the environmentalists, can have a larger piece.

Notes

1. For further development of this thesis, see Salinas-León (1993).
2. Another relevant and highly interesting issue, forcefully pointed out by Milton Friedman and James Buchanan on various occasions, is whether the establishment of a free-trade zone is a necessary condition for free trade. I concur with Friedman and others that unilateral opening is all that is required.

Indeed, as Brink Lindsey from the Cato Institute pointed out to me, the NAFTA could be contained in a single sheet of paper expressing the commitment of the three member nations to eliminate all barriers to trade. Instead, we have a 2,000-page document couched in highly complex legalese; see Salinas-León (1992c).

3. The first of such polls appeared in a leading Mexican magazine *Este Pais*, (April 1991). The latest comes from another leading weekly, *Epoca* (December 21, 1992), which reports a 74 percent public support. These numbers have become standard.

4. In fact, December 17, 1992, the date when the NAFTA was signed, is already considered an historic date for Mexico. The atmosphere during that day in Mexico City seemed to suggest a cause for national celebration.

5. This formulation is based on the text of a congressional testimony I delivered before the U.S. Congress in support of free trade and "fast-track," Subcommittee on Commerce, Consumer Protection, and Competitiveness, Committee on Energy and Commerce, U.S. House of Representatives, Washington, DC, May 15, 1991. The text was subsequently published (Salinas-León 1991a).

Of course, it goes without saying that such a view is hardly universal. The Washington, D.C.-based Economic Policy Institute has systematically attacked NAFTA, as have representatives of the "leftist" intelligentsia in Mexico. Jorge Castañeda, one of Mexico's leading social critics, claims that NAFTA involves an intolerably "high price to pay" for Mexico. See Salinas-León (1992f) for a rejoinder to Castañeda.

6. For a more thorough elaboration of these notions, see Salinas-León (1992d).

8. For more on the renowned Vaquerias project, see Salinas-León (1991d).

9. For more on the virtues and vices of Mexico's privatization program, see Salinas-León (1990) and (1992e).

10. I develop the role of Mexico in the EAI in Salinas-León (N.d.).

11. For a detailed assessment of the failings of Mexico's system of property rights, see Sarmiento (1992, A13); Damm (1991, 30–6); and Salinas-León (1991c, 8) and (1992b, 112–117).

12. For more on the special nature of Mexico's current and trade deficit, see Salinas-León (1992a).

13. Of course, there is far more behind the competitive forces that drive the flow of today's capital. See Clark and McKenzie (1991).

14. Thus, witness the recent restructuring of the state oil monopoly PEMEX (long-time sacred cow) and the recent decision to sell off secondary petrochemical operations, for an estimated worth of $6 billion. In addition, the Salinas government is currently allowing private provision of hydro-electrical services, as well as encouraging private capital participation in the state-owned Federal Electricity Commission. These two bureaucratic monstrosities will inevitably find themselves in a process of privatization, due in large part to the forces of unrestricted trade and its impact in framing a consistent regulatory regime.

15. I owe the basis of this argument to Lawrence H. Summers (1992, 299–300). See also Rubio (1991).

16. Notice, indeed, that market access merely builds upon a substantial reduction in tariffs under trade liberalization. Since 1986, Mexico slashed tariffs from an average of 80 to 10 percent and eliminated 96 percent of import licenses. The phase-out of zero tariff levels in a maximum period of fifteen years is seemingly conservative in comparison to the previous adjustment. For Mexico, therefore, there is more to the NAFTA than gaining broader market access.

17. Indeed, Mexican labor is far from cheap. If measured in terms of productivity output, it turns out that U.S. labor is less expensive. After all, what companies seek is not necessarily cheap labor, but productive labor. See Salinas-León (1991b).

18. See the already classic paper by Grossman and Krueger (1991).

References

Clark, Dwight, and Richard B. McKenzie. 1991. *Quicksilver Capital.* New York: The Free Press.

Damm, Arturo. 1991. *Liberalización en México.* Mexico City: Endomex.

Grossman, Gene, and Alan Krueger. 1991. Environmental impacts of a North American Free Trade Agreement. Presented at a conference on the U.S.–Mexico free trade agreement, Princeton University, October.

Rubio, Luis. 1991. *Acuerdo de Libre Comercio.* Mexico City: Editorial Diana.

Salinas-León, Roberto. 1990. Privatization in Mexico: Good, but not enough. *Backgrounder,* no. 797. Washington, DC: Heritage Foundation, November 15.

———. 1991a. A Mexican view of free-trade. *Foreign Policy Briefing,* no. 9. Washington, DC: The Cato Institute, May 21.

———. 1991b. Cheap labor and cheap arguments. *The News.* Mexico City, November 7.

———. 1991c. Legal reform and institutional reliability. *El Financiero International* (June 12): 8.

———. 1991d. Sweet land of liberty. *The News.* Mexico City, October 17.

———. 1992a. Don't cry for Mexico's current account deficit. *The Wall Street Journal* (February 21): A13.

———. 1992b. Economic reform in Mexico. In *The North American Free Trade Agreement: Spurring prosperity and stability in the Americas,* edited by Michael G. Wilson and Wesley R. Smith. Washington, DC: The Heritage Foundation, 112–117.

———. 1992c. El Regionalismo: un Obstáculo para el Tratado Trilateral. *Visión* (February).

———. 1992d. Independence and false nationalism. *The News.* Mexico City, September 12.

————. 1992e. Privatization in Mexico: Much better, but still not enough. *Backgrounder Update*, no. 172. Washington, DC: Heritage Foundation, January 20.

————. 1992f. The valid high price of NAFTA. *The News*. Mexico City, December 9.

————. 1993. North American free trade and the environment: The 'green' side of NAFTA. *Regulation* 16(1).

————. N.d. Mexico's role in the enterprise for the Americas. *Backgrounder*. Washington, DC: Heritage Foundation (forthcoming).

Sarmiento, Sergio. 1992. Seeking a legal complement to Mexico's opening market. *Wall Street Journal* (April 3): A13.

Summers, Lawrence H. 1992. Regionalism and the world trading system. *Policy implications of trade and currency zones*. Kansas City: Federal Reserve Bank of Kansas City.

3

The Environmental Impacts of Trade Liberalization

Steven Globerman*

Introduction

From the outset of negotiations between Canada, Mexico, and the United States to implement a North American Free Trade Agreement (NAFTA), environmental concerns featured prominently in the public policy debate. Specifically, opponents of the NAFTA argued that further trade liberalization, especially between Mexico and the United States, would result in significant incremental environmental damage. In order to win Congressional approval to negotiate a NAFTA along the "fast-track," the Bush administration agreed to carry on parallel negotiations concerning environmental issues alongside the trade negotiations, and President Clinton has said that these parallel negotiations are imperative.

Now that a NAFTA has been negotiated, the debate has become more sharply focused around the question of whether the specific agreement does enough to recognize environmental concerns (see Canada 1992). While government officials in the three countries have touted the NAFTA as being the "greenest" trade agreement ever produced, opponents of the NAFTA are claiming it does not go far enough to recognize and remediate the damaging effect that increased trade will have on the environment.

Specific environmental provisions in the NAFTA are similar to provisions

* The author thanks Brian Globerman and Daryl Madill for research assistance.

in the General Agreement on Tariffs and Trade (GATT). For example, the NAFTA allows specific environmental agreements among Canada, Mexico, and the United States to take precedence over the NAFTA provisions. The NAFTA affirms the right of each country to chose its own level of environmental protection. Moreover, each country may maintain and adopt standards and phytosanitary measures, including those more stringent than international standards, to secure its chosen level of protection.

There are provisions in the NAFTA which establish standards subcommittees to work to make compatible standards-related measures in specified areas including vehicle emissions and other motor carrier environmental pollution levels. The parties also agree to promote making compatible standards-related measures that are developed or maintained by state, provincial and local authorities and private sector organizations (Canada 1992, Annexes 913 A–C); however, there is nothing in the agreement which obliges countries with "stricter" environmental standards to harmonize their standards "downwards" to match those of their more "lax" trading partners. On the contrary, a country is free to raise its environmental standards to any level.

To be sure, disputes may arise over whether specific environmental provisions are merely disguised trade barriers. In disputes regarding a country's standards that raise factual issues concerning the environment, that country may choose to have the dispute submitted to the NAFTA dispute settlement procedure rather than to procedures under another trade agreement such as the GATT.

The same option is available for disputes concerning trade measures taken under specified international environmental agreements. The panel hearing the dispute will presumably seek to determine if the action taken is credible on environmental (or related) grounds or whether it is transparently a trade protectionist measure. In dispute settlement, the complaining country bears the burden of proving that another NAFTA country's environmental or health measure is inconsistent with the NAFTA.

In what is arguably a departure for an international trade agreement, the NAFTA contains general statements that the signatories will work jointly to enhance the protection of human, animal and plant life and health and the environment. The agreement also embodies a general statement that no NAFTA country should relax its health, safety, or environmental standards for the purpose of attracting or retaining investment in its territory. A party to the agreement who feels that another has offered such an encouragement may request consultations with the other party. If the two parties cannot resolve their dispute through consultation, the dispute can be sent to an arbitration tribunal. In short, the NAFTA embodies provisions to protect environmental amenities and ensure that these amenities are not sacrificed to attract investment.

The purpose of this chapter is to assess environmental criticisms of the NAFTA including the broad argument that stronger environmental provisions should be written into the current version of the NAFTA. The study proceeds by identifying the main potential interactions between trade liberalization and the environment and by setting out the important empirical relationships bearing upon the interactions. The available evidence surrounding the identified relationships is discussed in the third section. The fourth section considers a number of potential indirect linkages between trade patterns and the environment, including the potential for two-way interaction between trade liberalization and environmental protection. The chapter ends with a summary and some policy conclusions.

Economic Incentives and the Environment

While a host of seemingly heterogeneous concerns have been raised about the impacts of free trade, most can be related to, what economists call, substitution and income effects. In order to identify these effects in a relatively integrated fashion, it is useful to outline a conceptual model of the direct linkages between international trade and the environment.[1]

A Conceptual Model

Any economic activity can be characterized as a transformation of different quantities of inputs into quantities of outputs. For example, the production of a personal computer requires plastic, integrated circuit boards, machinery and labor. The production of a motoring vacation requires motor vehicle, fuel, and the time and expertise of the driver.

While economists have traditionally focused on labor and capital inputs, it is increasingly recognized that production activities involve the direct or indirect usage of the environment (see Dasgupta 1990). Hence the motoring vacation also generates a by-product of carbon monoxide fumes which affect air quality. The production of personal computers may result in an increase in solid waste material which must be stored in landfills that may create visual pollution or health hazards. Of course, some economic activities will use environmental inputs more intensively than others.[2] Indeed, even within any given economic activity, e.g., oil refining, there are likely to be more or less environmentally intensive production alternatives.

In this context, the environmental impact of any policy, such as trade liberalization, ultimately can be equated to the impact of that policy on the cost of environmental inputs. Two cost measures might be identified: (1) the change in the absolute cost of environmental inputs used and (2) the change in environmental intensity or the share of environmental costs in total costs. Presumably, individuals who believe any deterioration in the natural environment is unjustified will be concerned with absolute increases in environmental costs, whereas those more amenable to a trade-off between economic growth and environmental preservation may be more concerned with changes in relative environmental effects.

There are also alternative production processes available to produce the product, each process reflecting a different degree of substitutability among inputs including the environment. For example, the use of scrubbing and filtering equipment in plants represents an indirect substitution of capital for environmental inputs. The actual substitution among inputs will depend on, among other things, changes in the relative input prices reflecting scarcity, technology, and governmental policy. The relative prices of different inputs can vary both across producers and across geographic regions.

This framework points toward potential linkages between trade liberalization and environmental impacts. It suggests that the absolute level of environmental inputs used by the economy will be a function of:

1. the overall level of economic activity. All other things constant, the higher the overall level of production activity, the greater the absolute cost of environmental inputs.
2. the mix of output. Shifts in demand toward activities that are more environmentally intensive will, other things constant, result in a higher cost of environmental inputs.
3. changes in production technology. Technological change can reduce all inputs in proportion to one another (unbiased technological change), or it can reduce the use of one input relative to another (biased technological change). In either case, if technological change improves the efficiency with which environmental inputs can be used, it will reduce the environmental intensity of production.

Implementing the Framework

This conceptual framework can help identify the relationship between trade liberalization and the environment. Obviously, the impacts of trade liberalization on different aspects of economic performance will depend on the precise nature of any agreement. For our purposes, the most important feature of the NAFTA is its call for the ultimate removal of all tariff and nontariff border restrictions among the three countries. A related feature is that domestic environmental and health standards will continue to be set by each sovereign government, although national treatment will be accorded foreign companies.[3]

Increased demand. The elimination of trade barriers will influence the overall level of economic activity in the free trade area. Specifically, it will lead to higher income levels by promoting increased efficiency. Higher income levels, in turn, should encourage increased demand for goods and services with corresponding increases in consumption and production activities.[4]

While higher real incomes resulting from improved allocative and technical efficiency will increase overall demand, the impact this has on the environment will also depend on relative shifts in demand between products and relative shifts in production technologies. That is, income elasticities of demand will differ across the various products. Indeed, entirely new products might enter the economy at higher income levels. The demand for some products, especially those that are environmentally "dirty" or "unsafe" might actually decline. For example, used cars might be retired faster as people buy new, environmentally "cleaner" cars. As another example, people may build better insulated houses, thereby burning less wood or coal. To the extent that demand increases more rapidly for goods that are less environmentally intensive, the use of environmental inputs might actually decrease as total production increases. Whether an increase in overall demand increases the use of environmental inputs, is ultimately an empirical issue which will be reviewed later.

In the context of the NAFTA, available research suggests that real income effects will be small for the United States and Canada, whereas they will be absolutely and relatively large for Mexico.[5] This implies that the greatest impact on changes in North American consumption patterns will be associated with changes in Mexican real income levels.

Relocation of economic activity. The major concern of many environmentalists with respect to a NAFTA is that it will indirectly encourage an increase in the demand for environmentally intensive production activities.[6] To see this more clearly, imagine that there are two production activities, A and B, corresponding to the production of two products. Further assume that there are two ways to produce each product. The first technique, in each case, is relatively environmentally intensive. The second is relatively capital intensive. On average, product A is more environmentally intensive than product B. In country X, legislation severely restricts the ability of firms to use environmentally intensive techniques. In country Y, there are no meaningful restrictions.

Assume that tariff levels between the two countries are initially so high that no trade takes place. Presumably, producers of both products in country Y are using the environmentally intensive techniques relatively more than producers in country X, but this is particularly true in the case of product A. Now all barriers to trade are removed. Presumably, there is increased specialization of production such that production of product A, on net balance, shifts toward country Y and away from country X. That is, producers of product A in country X are attracted to Y by the ability to use the more environmentally intensive production technique. At the same time, there will be some shift, on balance, of product B production towards country Y. This, in turn, implies a reduction in the use of the environmentally intensive technique; however, given that B was not as environmentally intensive as A, a substantial portion of its production in country Y may have taken the form of the non-intensive technique anyway. On balance, demand for environmentally intensive techniques will increase relative to non-intensive techniques. Depending upon the nature of the reallocation of production activities, there could be an increase in the absolute and relative use of environmental inputs. For example, environmentalists argue that with its weaker record of environmental protection, Mexico will attract pollution-intensive activities that are increasingly uneconomical to carry out in North America. Within our conceptual framework, this is analogous to saying that specific production activities in Mexico will become cheaper under free trade and, therefore, that these activities will be in greater demand. At higher levels of intensity, they will lead to increased use of environmental inputs.

The relevance of the relocation argument therefore depends very much upon the impact of environmental standards on the choice of production technique. If the significance is limited, opening up the "production set" to include more environmentally intensive techniques may have little impact on producer behavior. If this opening of the production set is country specific, e.g., you can use environmentally intensive techniques in Mexico, the insignificance of environmental standards as a determinant of production techniques implies that they will have limited impacts on industrial relocation decisions.

Improved efficiency and increased transportation. Increased competition associated with trade liberalization should encourage improved technical efficiency as firms utilize optimally sized plants and specialize production within plants to realize economies of scale.[7] Increased competition should stimulate a faster rate of adoption of best practice technology, including technology associated with the direct or indirect use of environmental inputs.[8] On balance, this is likely to lead to a more efficient use of all inputs, including the environment.

Increased regional specialization could lead to increased intra-industry trade which might imply greater cross-border shipments of products Transportation tends to be relatively intensive in its use of environmental inputs. By itself, this should result in increased usage of environmental inputs; however, if there is a redirection of the flows of traffic and if different geographic regimes have different standards for vehicle emissions, it is conceivable that the total use of environmental inputs need not increase.

As with relocation, the use of environmental inputs tied to transportation activities will increase (decrease) if transportation increases (decreases) in "low standard" areas relative to "high standard" areas. For example, to the extent that the enforcement of environmental standards in Mexico is less stringent than in the United States, an increase in the relative amount of transportation between countries relative to within countries might actually reduce the environmental intensity of transportation activities.

The underlying logic of this assertion is that any vehicle travelling in the United States must meet U.S. standards, whether it is entering Mexico from the United States or entering the United States from Mexico. If it is increasingly profitable for vehicles to do relatively more travelling within the United States, it is increasingly profitable for vehicle owners to adopt less pollution intensive transportation activities. This substitution effect, of course, could be more than offset by an overall increase in transportation activities, both pollution and non-pollution intensive, tied to higher levels of economic activity, as well as longer distances travelled, on average. The implication here is that the environmental impact of trade liberalization depends, among other things, upon the precise response in terms of transportation activities.[9]

Technological change. Changes in technology may permanently alter the input structure of consumption or production activities. For example, improved heat recirculation and monitoring systems may reduce the demand for energy inputs in residential and commercial space heating such that there is a permanent substitution of capital for energy at all possible relative prices of these inputs. Free trade can be expected to speed up the adoption of new technology; however, it is not obvious that it will bias technological change towards or away from the use of environmental inputs.[10]

Empirical Evidence Linking Trade and the Environment

In this section, we consider some available evidence on several of the important relationships identified in the preceding section. One important relationship is the impact of higher income levels on the demand for different economic activities. Most directly, the income elasticities of demand for environmentally intensive activities may differ from those for non-intensive activities. Hence, as income levels rise, consumption may increase for one set of activities relative to another, possibly resulting in lower absolute levels of environmental input usage.

Less directly, if a clean environment is itself an income elastic good, the legislating and enforcing of environmental standards may become stricter as a country's income level increases. In effect, the relative price of environmentally intensive production activities will increase leading to a substitution in favor of non-intensive production activities.

The foregoing discussion suggests that the relationship between real income levels and the use of environmental inputs may not be linear. That is, over an initial range of income, the dominating influence will be the overall level of economic activity which encourages environmental input usage, all else constant. But beyond some point, the changing mix of activities towards less polluting ones will come to be the dominating influence, and environmental input usage may decline. The "switchover point" is ultimately an empirical issue.

Evidence on Income Effects

In the absence of accurate forecasts of real income changes and the income elasticities of demand for different products (including environmental amenities, per se), it is impossible to be precise about these direct and indirect consumption effects; however, it can be reasonably asserted that as Mexico's real income level increases, consumption patterns will at some point increasingly favor non-environmentally intensive activities. In particular, demand for services will grow relative to demand for manufactured goods. The former are less environmentally intensive than the latter. Moreover, demand for health, visual amenities such as landscape views, recreational amenities involving relatively clean water and so forth are highly income elastic.

Some available studies provide support for the hypothesis that the demand for a cleaner and healthier environment is strongly and positively related to higher real income levels, at least beyond some threshold income level. For example, Grossman and Krueger (1991) correlated the level of sulfur dioxide and smoke with per capita income and found that the level of pollution rises until income reaches $5,000 per head (in 1988 dollars) and then starts to fall.[11] Mexico's real income level per capita in 1991, measured as gross domestic product per capita in U.S. dollars, at $2,365 was below the $5,000 threshold. By itself, this suggests the potential for the mix of economic activities in North America to become more environmentally intensive in the early years of a NAFTA; however, the continued economic growth of Mexico associated with the NAFTA would ultimately reverse this conclusion. That is, to the extent that the NAFTA accelerates the growth of the Mexican economy, it will lead to less sulfur dioxide pollution in the long-run, since it will shorten the time it takes Mexico to reach the "crossover" income level.

It might be noted that estimates by the World Bank also indicate a curvilinear relationship between the average ambient level of sulfur dioxide and real income per capita; however, the World Bank places the switchover income level at closer to $2,500 (see *Economist* 1992, 8). This latter estimate would suggest that the effects of the NAFTA are likely to be benign in both the short and long run, since Mexico will cross this income threshold in the near future with or without a NAFTA in place.

Some additional evidence on the relationship between income levels and environmental amenities is provided in a study by Walter and Ugelow (1978). Based on questionnaires sent to national officials in developed and developing countries, they found that while the strictness of environmental policies varied within each group, the level of strictness was nonetheless higher, on average, in developed countries. This finding is also consistent with observations that urban sanitation tends to be an increasing function of income at all income levels, while ambient levels of particles tend to be a decreasing function of income at virtually all income levels (see *Economist* 1992). This latter observation suggests that the NAFTA will unambiguously reduce pollution related to sewage and ambient particles to the extent that it accelerates income growth in Mexico and, to a lesser extent, in Canada and the United States.

To be sure, some forms of environmental pollution increase with higher national income levels. For example, carbon dioxide emissions tend to increase fairly uniformly with higher income levels, as does solid waste (*Economist* 1992). These observations qualify an unambiguous conclusion that the income effects of a NAFTA will either be benign or favorable for environmental amenities in North America. Nevertheless, taken on balance, one must conclude that economic growth stimulated by a NAFTA may well be positive, on balance, for the environment in terms of reducing the utilization of environmental amenities. The main effect here is the increased demand for a cleaner environment which is associated with a shifting away from pollution intensive activities.[12]

Evidence on Relocation Effects

There is a fairly substantial body of evidence bearing upon the relocation effect of trade liberalization. The evidence, by and large, suggests that the effect is likely to be very modest and restricted to a few industries.[13] Given the importance of this phenomenon in the debate surrounding the NAFTA, it is worth reviewing the available evidence in detail.

In one study, Leonard (1984) found no evidence in overall statistics on foreign investments by U.S. companies and U.S. imports of manufactured goods that key high pollution industries have shifted more production facilities overseas in response to environmental regulations. Yet in a few high pollution, hazardous production industries, environmental regulations and work place health standards have become a more prominent and possibly decisive factor in industrial location and have led U.S. firms to move production abroad. Examples of such industries are those that produce highly toxic, dangerous or carcinogenic products, such as copper, zinc and lead. For these latter industries, environmental regulations have combined with other changing locational incentives and economic problems to speed international dispersion of capacity.

In a similar vein, Walter (1982) reports that certain copper smelters, petroleum refineries, asbestos plants and ferroalloy plants have reportedly been constructed abroad rather than in the United States for environmental reasons. There has also been some resiting of petrochemical complexes and chemical plants within Europe for environmental reasons. Moreover, some recent Japanese pollution-intensive industries have reportedly been channeled to developing countries in Southeast Asia and Latin America; however, there is no evidence of a "massive" environment-induced locational shifting of production capacity. Moreover, a significant amount of the observed geographical mobility of production involves cases where major projects were absolutely barred for environmental reasons.

Rubin and Graham (1982) conclude that during the decade of the 1970s, when complex environmental regulations and high pollution costs were imposed on industries, the overall foreign investment and import trends of the mineral processing, chemical and pulp and paper industries did not differ fundamentally from those of U.S. manufacturing industries in general. The former industries are arguably among those that should have been most adversely affected by environmental legislation implemented in the United States. In fact, only slight shifts at the margin could be detected in these industries. Specifically, only a few U.S. industries within branches of the chemical manufacturing sector have increased production overseas as a direct or indirect result of environmental regulations. Two reasons are offered for this minor effect: (1) environmental control costs are a relatively small proportion of total production costs, and (2) most industries that have been hit hard by environmental regulations have been able to adapt to them by changing their production processes or by using different raw materials.

Stafford (1985) examined whether traditional factors such as access to markets, and differences in costs of labor and materials remain predominant in manufacturing location decision making, despite the recently added dimension of environmental regulations introduced under the National Environmental Policy Act. Specifically, personal interviews and mailed questionnaires were used to identify the factors that were most important in the location of 162 new branch plants of U.S. corporations. For most of the locational decisions investigated, environmental regulations did not rank among the most important factors considered. When such regulations were of some significance, uncertainties about when the necessary permits would be obtained were more important than spatial variations in direct cost. Stafford (1985) concludes that environmental regulations have had no consistent effect on the size of the search area, the number of sites considered, the sizes of facilities built or the decision to expand existing plants versus building new plants.

Bartik (1988) used a database of new manufacturing branch plants opened by Fortune 500 companies between 1972 and 1978 to determine if business location decisions are affected substantially by state environmental regulations. Two measures of state water pollution regulations and four measures of state air pollution regulations were used as variables. The study did not find any statistically significant effect of state environmental regulations on the location of new branch plants. Even sizeable increases in the stringency of state environmental regulations were found unlikely to have a large effect on location decisions for the average industry; however, for some highly polluting industries, the results cannot rule out the possibility of effects of environmental regulation on plant location.

Finally, McConnell and Schwab (1990) estimated a statistical model to investigate the impact of a variety of country characteristics on the locations of 50 new branch plants in the motor vehicle industry during the period 1973–1982, a period when there were wide variations in environmental regulations among regions. Most of the results indicate that environmental regulations do not exert an important influence on location decisions. At the margin, however, there is some evidence that firms may be deterred from locating where the ozone problem is severe and emission controls are correspondingly stringent.

In summary, the conclusion one might seemingly draw from existing studies is that geographic differences in environmental standards have a relatively small impact on the location decisions of firms. Indeed, any significant impacts appear concentrated in resource-based sectors that are arguably relocating from the United States for other reasons. Moreover, if these activities did not migrate to Mexico, they might migrate to other locations in the hemisphere where environmental standards are even lower. In short, the relocation criticism of the NAFTA seems quite overstated in light of the small share of costs in most industries ascribable to pollution abatement and the already low levels of U.S. tariffs in industries facing high pollution abatement costs (see Office of the U.S. Trade Representative 1991).

With respect to the linkage between trade liberalization and transportation patterns, it can be expected that cross-border traffic will increase with increased international trade. Indeed, congestion along the Mexican–U.S. border at major crossings in Texas and California is seen as being suggestive of the even larger problems that will emerge under a free trade agreement. Without gainsaying the wisdom of improving the transportation arteries between the United States and Mexico, it is unclear whether additional transportation implies significant increases in vehicle emissions. As noted earlier, an important factor in this regard is what happens to Mexican transportation vehicles. Specifically, to the extent that there is a decline in the average age of these vehicles, along with improvements in emission control equipment, increases in transport mileage need not translate into equivalent increases in vehicle emissions.

Trade Agreements and Environmental Legislation

Trade negotiations leading to trade liberalization agreements can have a potentially important indirect effect on environmental conditions by influencing environmental policies in the trading partners. In this regard, a concern raised by environmentalists is that increased competition associated with trade liberalization will lead domestic producers to "cheat" with respect to obeying environmental standards. Alternatively, or in addition, it will lead to increased lobbying efforts to have environmental standards relaxed (see Emerson and Mikesell 1991).

The argument that firms facing the competitive pressures of free trade will abandon environmental responsibility and ignore codified (or uncodified) standards, i.e., use illegal, pollution intensive production techniques, begs the question why they would not have cheated prior to the implementation of free trade if they thought they could do so with impunity. Perhaps the risks of getting caught are worth taking when one is faced with the prospect of bankruptcy; however, massive increases in bankruptcy risks cannot be realistically contemplated as a possible consequence of a NAFTA.

A more plausible scenario is that national governments will be less inclined to pass and enforce environmental standards given industrial dislocations and short-term unemployment associated with adjustments to trade liberalization. Indeed, governments might rely upon reduced regulation of business as a form of "adjustment assistance" for domestic industries. A related possibility is that governments will lower domestic environmental standards in order to permiᵥ domestic firms to compete on a "level playing field" with firms based in countries with "lower" environmental standards or enforcement practices.

The assertion that trade liberalization leads to a de facto harmonization of environmental standards at the "lowest common denominator" is not supported by the evidence. For example, United States and Canadian tariffs have been reduced over time to imports from a wide range of countries having weaker environmental standards, yet there has been no retraction of standards or enforcement based upon these imports. As another example, by mid-1991, the European Community had adopted numerous directives and regulations dealing with environmental matters along with liability standards in cases involving pollution. While many member countries were unaggressive in enforcing the European Community's environmental rules, pressure from those countries enforcing the rules and adopting their own tough antipollution laws is apparently bringing about compliance by all members (see Portaria 1991, 52–3).

Environmentalists argue that the trade agreement should be used as a lever by Canada and the United States to extract stronger environmental commitments on the part of Mexico. But we must recognize the potential for the social costs of pollution control to exceed the social benefits. If so, pressure to extract higher standards from lower income countries raises the prospect of excessive reductions in the use of environmental inputs from an overall welfare economics perspective.[14] A more flexible position on the part of environmentalists might hold that cross-border pollution is intrinsically a matter of bargaining between countries with strict and relaxed environmental regimes and that harmonization of regimes should fully and fairly acknowledge cross-border environmental spillovers.

The issue in the context of this report is whether trade liberalization makes it more or less likely that cross-border pollution externalities will be recognized and effectively dealt with by the countries involved. Alternatively, should trade liberalization be made contingent upon the trading partners dealing with cross-border pollution? Unfortunately, there is no unequivocal theoretical answer to this question. That is, arguments can be made pro and con the position that environmental and trade negotiations should be directly linked in order to ensure that international pollution spillovers are addressed by polluters.[15]

What can be theoretically established is that it is rarely welfare improving for countries to impose trade restrictions in response to their being "polluted upon" by another country's producers or consumers (Lloyd 1992). This observation, combined with the recognition that lobby groups will use environmental issues to extract protection against imports, suggests the wisdom, on balance, of separating the treatment of environmental problems from the treatment of trade issues.

Experience does suggest that the negotiation of environmental standards is likely to be a difficult and costly process. For example, evidence from the European Community (EC) experience supports this perception, especially attempts to establish auto emission standards. In the latter case, France has historically argued for lower exhaust emission standards, whereas countries such as Holland, Denmark and Greece have argued for higher standards (see *Wall Street Journal* 1988; also see Cairncross 1990). However, if the EC experience attests to the difficulties in harmonizing environmental standards across trading partners, it also suggests that a convergence of standards will ultimately take place in the direction of the more restrictive set of standards as noted above.

In summary, closer linking of trade and environmental negotiations might lead to improved cooperation among trading partners in addressing problems of transborder pollution. On the other hand, any such linkage imposes a substantial risk that environmental concerns will be used by lobbyists to gain protection from imports. Indeed, there is an added risk that the trade liberalization process will unravel with the resulting costs of protectionism far exceeding whatever direct or indirect environmental benefits might have resulted.

In the last analysis, the level of intergovernmental cooperation to address border pollution problems is the single most important factor affecting environmental conditions in the member countries (see Office of the U.S. Trade Representative 1991). A contentious international trade environment is unlikely to promote an atmosphere of cooperation for intergovernmental agreements to address cross-border pollution. Furthermore, the expertise needed to address trade issues does not necessarily overlap the expertise needed to negotiate international environmental agreements.

If attempts by developed countries to enforce stricter standards in developing countries through withholding trade liberalization leads to a breakdown of the liberalizing process, it could also have significant indirect adverse consequences for environmental amenities. As noted earlier, trade liberalization in itself may encourage a more "enlightened" environmental stance on the part of weak enforcement regimes to the extent that it encourages more economical pricing of natural resources. In many developing countries, in particular, deliberate "underpricing" of petroleum products, chemical fertilizers and the like encourages the adoption of environmentally intensive production techniques.[16] A move towards marketization of economic activity, which is often a concomitant of trade liberalization, could encourage a marked reduction in the relative use of environmental inputs by promoting a more economically rational set of prices for inputs that heavily pollute the environment.

As well, the NAFTA as negotiated should lead to some geographical dispersion of production capacity away from the Mexico–U.S. border. Under the relevant legislation, Maquila plants are encouraged to locate along the border. The crowding along the border is arguably exacerbating pollution problems there, since the natural environment has a limited capacity to absorb and process waste. If free trade between Mexico and the United States was derailed by environmental complaints, the benefits of whatever geographical dispersion would take place will be obviated.

Summary and Conclusions

Environmental criticisms of the NAFTA assume that trade liberalization inevitably leads to further environmental degradation; however, relatively simple modelling of the relationship between the trade and the environment suggests that the relationship is complex, and even the direction of the relationship is uncertain. Certainly, one can make a credible argument that, in the long run, free trade will lead to less intensive use of environmental inputs, both in relative and absolute terms. The primary source of this result is the certainty that, as countries become wealthier, an increasing demand for a cleaner environment eventually leads societies to choose less environmentally intensive production and consumption activities.

Linking trade liberalization to the environmental policies implemented by a country's trading partners can arguably be an instrument to "encourage" relatively lax countries to impose and enforce tighter environmental standards. Whether wealthy countries should "dictate" policy for purely local environmental problems is beyond the scope of this study, as is the issue of whether developed countries should pay developing countries to cease certain practices which are offensive to environmentalists from developed countries. What must be acknowledged is that a linkage can potentially focus the recalcitrant countries' minds on dealing with cross-border pollution.

At the same time, linking trade law to environmental policy invites a risk that protectionist forces will seize upon differences in environmental standards as an excuse to trigger countervail duties, anti-dumping penalties and other nontariff barriers to imports. It certainly invites a risk that the trade liberalization process will be derailed. The resulting consequences are certain to be adverse for both the developed and the developing country.

Environmentalists have implicitly or explicitly expressed skepticism about trilateral negotiations focused directly on environmental issues producing meaningful results; i.e., leading to commitments to deal with transborder pollution problems. Experience from the EC suggests that negotiations surrounding the harmonization of environmental standards does not necessarily lead to harmonization around the lowest standards. Indeed, it is not clear that linking trade liberalization to the negotiation of environmental standards leads to a harmonization of standards at a higher level than would otherwise be the case. There is a risk that the underlying trade liberalization process will be curtailed and with it a growing economic interdependence among the trading partners. Economic interdependence is arguably the most important factor, in the long run, leading to cooperation on other matters among governments.

In the final analysis, a failure of the NAFTA negotiations could lead to increased pollution for at least two possible reasons: (1) it could lead to the disruption of growing cooperation between the U.S. and Mexican governments to address border pollution issues;[17] and (2) without a NAFTA, increased emphasis might be placed on the maquiladora sector with even greater environmental stresses being placed on U.S.–Mexico border regions. This latter observation suggests that the nature, as well as the extent, of economic integration can affect the important linkage between economic growth and the utilization of environmental inputs.

Notes

1. A more formal development of this framework is available from the author on request.

2. It should be explicitly acknowledged that, in many instances, producers and consumers will utilize environmental inputs without incurring any personal explicit costs. Rather, the cost is borne as an externality by others. In this analysis, costs of environmental inputs are viewed in the broad social sense; i.e., they reflect the opportunity cost of the environmental resources "used up" regardless of whether an explicit pecuniary cost is imposed upon sets of producers or consumers.

3. National treatment implies that a single set of domestic environment and health and safety standards will be applied for foreign and domestically owned firms. In accordance with its commitment to Congress to obtain fast-track authorization, the U.S. government informed Canada and Mexico that in a NAFTA it must maintain the right to prohibit the entry of goods that do not meet U.S. health, safety, pesticide, food and drug and environmental regulations, so long as such regulations are based on sound science, do not arbitrarily discriminate against imports or constitute a "disguised" trade barrier. See Office of the U.S. Trade Representative (October 1991, 2).

4. If trade diversion discourages imports from nonmember countries, there could be a decline in levels of economic activity outside North America.

5. For a review of this evidence, see Globerman (1992).

6. For example, Shrybman (1991, 27) argues that: "One way in which the principles of free trade or deregulated trade have operated to undermine environmental regulation is by making it easier for corporations to establish or relocate operations to jurisdictions where the cost of doing business, including the cost of environmental regulation, is lowest."

7. Such rationalization is a ubiquitous response to trade liberalization agreements. For a comprehensive review of the evidence, see Globerman (1988).

8. Similar points are made in a GATT report. Specifically, the report notes that the opportunities for countries to trade in world markets for technology

facilitates the implementation of needed environment-improving processes. Similarly, trade can help consumers make environmentally beneficial choices—for instance, imports of low sulphur coal can encourage them to abandon the use of polluting high sulphur coal. See General Agreement on Tariffs and Trade (1992).

9. It also obviously depends upon whether there is a harmonization of allowable emission levels and other vehicle related pollutants as part of the NAFTA, either towards more or less restrictive standards.

10. It can be argued, however, that market reforms associated with trade liberalization might lead to the reduction of domestic subsidies. In developing countries, including Mexico, energy inputs and agricultural chemicals are frequently subsidized. The elimination of these subsidies will arguably lead to reduced use of these environmentally intensive inputs and, presumably, to less disruption of environmental amenities at all income levels.

11. Also see studies in Braden and Kolstad (1991).

12. Note that such shifting can also reflect greater utilization of pollution abatement equipment in activities that were relatively pollution intensive.

13. This is also the conclusion of a comprehensive review of the environmental economics literature which ascribes this result to the fact that the costs of pollution control not, in fact, loomed very large even in heavily polluting industries, i.e., on the order of only 1 to 2.5 percent of total costs in most pollution-intensive industries. See Cropper and Oates (1992).

14. The economically efficient set of environmental standards, i.e., the set that leads to the use of environmental inputs up to the point where the marginal social benefit equals the marginal social cost will vary from country to country depending upon a host of factors including topography, climate, demography and so forth. For a discussion of this issue, see Pearson (1987).

15. For a discussion of these points, see Lloyd (1992).

16. See Anderson and Blackhurst (1992). In the case of the NAFTA, freely traded energy products might also encourage Mexican energy users to buy cleaner fuels such as natural gas and energy from the United States to replace the use of high sulphur fuel oil.

17. In some models, the level of intergovernmental cooperation to address border pollution is the single most important factor affecting environmental conditions in the member countries. See Office of the U.S. Trade Representative (1991).

References

Anderson, Kym, and Richard Blackhurst. 1992. Trade, the environment and public policy. In *The greening of world trade issues*, edited by Kym Anderson and Richard Blackhurst. New York: Harvester Wheatsheaf, 3–22.
Bartik, Timothy. 1988. The effects of environmental regulation on business

location in the United States. *Growth and Change* 19:22–44.

Braden, J. B., and C. D. Kolstad, eds. 1991. *Measuring the demand for environmental quality*. Amsterdam: North-Holland.

Canada. 1992. *North American Free Trade Agreement*, draft legal text, September 8.

Cairncross, Frances. 1990. Environmental barriers going up as EC takes trade barriers down. *Financier* 14:16–20.

Cropper, Maureen L., and Wallace Oates. 1992. Environmental economics: A survey. *The Journal of Economic Literature* 30 (June): 675–740.

Dasgupta, Partha S. 1990. The environment as a commodity. *Oxford Review of Economic Policy* 6:1–18.

Economist. 1992. The environment: Whose world is it, anyway? 323(7761).

Emerson, Peter, and Raymond Mikesell. 1991. North American free trade: A survey of environmental concerns. *Policy Briefing*. San Francisco: Pacific Research Institute, December.

General Agreement on Tariffs and Trade. 1992. Expanding trade can help solve environmental problems, press release. Geneva, February.

Globerman, Steven. 1988. *Trade liberalization and imperfectly competitive industries: An overview of theory and evidence*. Ottawa: Economic Council of Canada.

———. 1992. The economic integration of North America. Paper presented at the general meeting of the Mont Pelerin Society, August 30–September 4, at Vancouver, British Columbia, Canada.

Grossman, Gene, and Alan Krueger. 1991. Environmental impacts of a North American free trade agreement. Presented at a conference on the U.S.–Mexico free trade agreement, Princeton University, October.

Leonard, H. Jeffrey. 1984. *Are environmental regulations driving U.S. industries overseas?* Washington, DC: The Conservation Foundation.

Lloyd, Peter J. 1992. The problem of optimal environmental policy choice. In *The greening of world trade issues*, edited by Kym Anderson and Richard Blackhurst. New York: Harvester Wheatsheaf, 49–92.

McConnell, Virginia, and Robert Schwab. 1990. The impact of environmental regulation on industry location decisions: The motor vehicle industry. *Land Economics* 66:67–81.

Office of U.S. Trade Representative, Interagency Task Force. 1991. *Review of U.S.–Mexico environmental issues*. Washington, DC, October.

Pearson, Charles S. 1987. Environmental standards, industrial relocation and pollution havens. In *Multinational corporations, environment and the Third World: Business matters*, edited by Charles S. Pearson. Durham: Duke University Press, 113–28.

Portaria, Rudi. 1991. Toughened environmental regulation looms in EC. *National Underwriter* 95:52–3.

Rubin, Seymour, and Thomas Graham. 1982. The relation of international trade and environmental policy. In *Environment and trade*, edited by Seymour

Rubin and Thomas R. Graham. Totawa, NJ: Allanheld, Osmun and Co.

Shrybman, Steven. 1991. The environment and free trade. *Briarpatch* (September): 26–8.

Stafford, Howard A. 1985. Environmental protection and industrial location. *Annals of the Association of American Geographers* 75:227–40.

Wall Street Journal. 1988. France's switch on auto-exhaust curbs shows difficulty of unifying EC rules. (August 10): 13.

Walter, Ingo. 1982. International economic repercussions of environmental policy: An economist's perspective. In *Environment and trade*, edited by Seymour Rubin and Thomas R. Graham. Totawa, NJ: Allanheld, Osmun and Co.

Walter, Ingo, and J. Ugelow. 1978. Environmental attitudes in developing countries. *Resources Policy* 4:200–9.

4

The Environmental Side of North American Free Trade

Peter M. Emerson and Robert A. Collinge

Introduction

Trade between the countries of North America accounts for nearly one-fourth of all world trade. It affects the scale and composition of economic activity at home and abroad. Reflecting the decisions of millions of individual producers and consumers, this economic activity generates wealth and contributes to environmental degradation. Most trade and financial transactions in North America occur between the highly-developed economies of the United States and Canada, which signed a free trade agreement in 1989. The United States and Canada also share environmental and natural resource problems, and have entered into several international agreements to help solve them.

In 1987, Mexico joined the General Agreement on Tariffs and Trade (GATT). Since then, trade barriers have been reduced and trade between the United States and Mexico has grown significantly. In 1990, President Salinas approached President Bush to negotiate a free trade agreement with Mexico. A few months later, Canada joined the negotiations and the three countries began working for a North American Free Trade Agreement (NAFTA).[1]

The proposed NAFTA has stimulated a vigorous debate about the benefits and costs of such an agreement, especially as they relate to the environment, public health, and industrial location and employment. The stimuli for this debate most likely include the following factors: wide differentials in the standards of living between the countries; a weak economy in the United States and Canada; growing awareness of environmental and public health problems along the United States–Mexico border; and Mexico City's unfortunate status as the world's worst metropolitan air polluter.

This chapter first examines the nexus between trade policy and environmental protection in North America. While there are many unanswered questions, there is little doubt that the environment will be one of the most important trade issues of the 1990s. The questions that will be debated in the trade context are: to what extent does freer international trade exacerbate or ameliorate environmental problems; and what steps, if any, must be taken to reduce the negative impact of trade on the environment?

From a U.S. perspective, the chapter then discusses the environmental aspects of the proposed NAFTA and several options that remain as the trade agreement is taken up by the Clinton Administration and a new Congress. Environmentalists certainly do not want the trade agreement to weaken environmental laws. Instead, at least some environmentalists look to freer trade and economic growth as a means of improving environmental protection for all people (Emerson 1991; Hair 1991; Goebel 1992). It seems likely that a lasting effect of the NAFTA debate will be a "greening" of future trade pacts.

Freer Trade and Environmental Protection: Are They Compatible?

Free trade in its ideal sense means that governments place no restrictions on the transfer of goods, services, capital and other resources across national boundaries. In this ideal world, wealth is increased and efficiency enhanced as long as the producers of the goods and services must pay the full cost of production, including the use of the environment as a receptacle for waste products. When these environmental costs are not fully accounted for by producers, however, governmental regulations may be required to protect the environment and public health. The question then arises as to whether free trade and environmental protection can complement each other and promote public welfare (Reinstein 1992; U.S. Congress 1992).

International trade agreements, such as the NAFTA, provide detailed rules aimed at reducing barriers to commerce, settling disputes among trading partners and otherwise regulating trade. Given political factors and other constraints, the objective of trade negotiators is to move the countries in the direction of freer trade.

The NAFTA does not liberalize trade very much, since trade barriers between the countries of North American already have been reduced. However, even without a dramatic lowering of barriers, it may still have a pronounced effect in terms of promoting economic activity. The reason is that the agreement adds certainty. It provides a guarantee that lower trade barriers and other economic reforms being pursued in Mexico will continue beyond the tenure of current politicians or political parties. The long-term planning this allows serves to hasten adjustment by businesses on both sides of the border, resulting in more investment and increased trade. Thus, even if the agreement does not result in large reductions in trade barriers, it stimulates international business by offering some assurance that current policies toward trade and investment will continue into the future.

The case for freer trade is based on the argument that trade according to each country's comparative advantage increases the total value of output in the trading countries. Unrestricted prices in competitive markets achieve this result by signaling each country to produce those goods for which it has the lowest opportunity cost, and to trade for the rest. In the short run, there will be winners and losers; but ultimately, efficient producers prosper and consumers receive larger volumes of goods and services at lower prices (Collinge and Emerson 1992).

Unfortunately, free trade does not lead to optimal trade patterns if there are externalities—such as transborder pollution damages—that are not corrected through public policy. In the context of trade between the United States and Mexico, these externalities may be significant. Specifically, the danger is that some of the environmental costs from production in one country, say Mexico, may be borne by "third party" residents in another country (the United States). Without a mechanism to force producers to bear these costs, there is little incentive to reduce pollution, and there is an inefficient advantage to polluting industries in Mexico over less-polluting competitors in other countries. In this context, policy measures are needed to "internalize" the transborder damages.

Relative to the status quo, freer trade within the context of the NAFTA may prove beneficial to the environment. One reason has to do with the existing maquiladora program, which allows factories located on the Mexican side of the border to trade freely with the United States, under provisions that products must be sold in the United States. The NAFTA in effect broadens the scope of the maquiladora program to include all of Mexico and eliminates the requirement that goods must be exported to the United States. Lifting these constraints is likely to prompt some of the maquiladora plants to move to the interior of Mexico, thus reducing pollution along the Rio Grande and other border areas. By itself, the NAFTA is unlikely to attract many new factories to border areas, since most factories that would find border locations attractive would be oriented toward the U.S. market and already had access to free trade under the maquiladora program. Hence, they are likely to be located along the border already.

Although environmental regulations in the United States and Mexico may be comparable, enforcement is often lax in Mexico. For instance, much of the hazardous waste generated in Mexico is thought to be dumped illegally (Kleist 1992). Lower levels of environmental quality in Mexico can be explained in large part by the lower per capita income of Mexican residents. Environmental quality is what economists call a normal good, meaning that as incomes rise, people will be willing to devote more income to obtaining a cleaner environment. This demand will be manifested in a variety of ways including tougher regulatory enforcement, for which citizens pay, in part, through higher prices resulting from increased production costs.

Environmental damages can be divided into two overlapping categories: (1) direct externalities resulting in first-order damages and (2) indirect externalities resulting in second-order damages. First-order damages represent the value of losses occurring to individuals, plants and wildlife, buildings, and other man-made and natural assets as a result of direct exposure to a pollutant or a destructive production or consumption process. Examples could include birth defects caused by a toxic emission, the loss of species due to deforestation, and illness caused by the release of raw sewage. Residents breathing the smog of Los Angeles or Mexico City are victims of first-order damages.

Second-order damages go beyond direct exposure–response impacts to account for losses which occur to individuals because they value the environment beyond their own physical proximity, or because they are concerned for the well-being of others. For instance, some people feel a loss if rain forests are logged or if the health of residents in Mexico City, Los Angeles or elsewhere is diminished. The cost of pollution and environmental degradation is found by summing the value of first-order damages and second-order damages.[2]

Whenever such environmental costs are ignored, polluting activities receive an implicit subsidy and consumers have too little environmental quality. For instance, residents in proximity to the border complain of various transborder damages. Maquiladora plants—often owned by U.S. firms—pollute the border environment. Untreated sewage from border communities flows into the Rio Grande, resulting in first-order damages to downstream users, both Mexicans and Americans.

This outcome leads to the possibility that the introduction of environmental regulation (i.e., an effort to internalize environmental costs) may change international trade, but enhance social welfare by removing the implicit subsidy. Specifically, any damage to the U.S. (Mexican) environment from pollutants originating in Mexico (United States) represents an external cost to the United States (Mexico) of that production. Decision makers in either country cannot be expected to consider these costs unless international agreements and policies induce them to do so.

The regulatory task is complicated by two factors. First, since the domain of environmental damages may be either domestic, transborder, or global, uniform standards across regional and national borders may not be justified. Second, the optimal amount of environmental quality may differ significantly between trading partners because of differences in preferences, income, and assimilative capacities.

For example, transborder pollutants aside, first-order damages fall on domestic residents, and provide the primary impetus for domestic policy action. The Mexican government's decision to restrict taxicab emissions and industrial output in Mexico City was based on first-order externalities. Relative to citizens of other countries, domestic residents are far more affected by their own first-order pollution, and therefore are in a better position to determine whether they are willing to pay the price for a cleaner environment. Likewise, pollutants emitted in Mexico will generally have their strongest physical impacts within that country.

Consider the second-order damages associated with the genuine concern of United States and Canadian citizens over the welfare of Mexican residents who must live with unhealthy levels of pollution, or the concern of all people for rainforest preservation. Rather than being a concern over pollution or environmental degradation per se, this is a concern over the welfare of current and future generations of people as it is impacted by broad concepts of ethics and equity. Thus, the concern is really over the standard of living in all its facets.

As noted above, the greater levels of pollution in Mexico can be traced to the lower Mexican per capita income that forces Mexicans to choose lesser amounts of all goods, including environmental quality. This suggests that the second-order externality of interdependent utility functions is best addressed by promoting economic growth, rather than through targeting pollution policies directly. The NAFTA offers a way to promote economic growth, which in turn is likely to increase the willingness and ability of Mexicans to pay to enforce pollution controls in their own country.

Some people are concerned that when production costs are not equal between countries, the country with the lower costs will attract industry from higher cost countries. The specter of large scale unemployment in the country opting for higher quality environmental and labor standards motivates these well-meaning people to oppose free trade. They do not wish to reward a country for choosing to become a "haven" for pollution or labor exploitation. Fortunately, while some migration of industry to "pollution havens" might occur, the effect on domestic output and employment is likely to be negligible.

There are two reasons that pollution havens have not resulted in large scale job losses. First, for many production processes, environmental costs are only a small portion of the firm's total costs and research shows that environmental factors have not been major determinants of how companies allocate their investments among countries (Pearson 1987, 124). The second reason revolves around the nature of trade itself. No country sells its products to another country unless it can receive something just as valuable in return. For instance, a Mexican textile maker might sell blankets in the United States in exchange for dollars. Those dollars are then respent on United States products either by the textile maker or, through the foreign exchange market, by someone else.

The foreign exchange market between the United States and Mexico can be thought of as equilibrating the desires of Mexicans to purchase U.S. goods, services, and investments with the corresponding desires of U.S. citizens to purchase Mexican goods, services and investments. In effect, dollars and pesos never leave their home countries. They head for the border but, through the foreign exchange market, bounce right back home. The result is that job losses in some sectors are offset by job gains in other sectors. Therefore, while this section has identified some legitimate concerns over different levels of environmental protection among the NAFTA signatories, the concern regarding job migration to pollution havens is not paramount. There are valid reasons for allowing environmental quality to vary from place to place. To some extent, concerns over differing environmental standards can be traced back to comparative advantage. At the current stage of development, the opportunity cost of providing environmental quality is higher in Mexico than in the United States and Canada. In the developed nation, citizens have higher income, and thus choose to consume a higher level of environmental quality along with more and better housing, food, clothing, and all other normal goods. As Mexican incomes rise, we will see the same pattern emerge in that country. In the interim, production costs are lower as workers are willing to provide their labor in exchange for less of all these things. To require Mexican citizens to support the same level of environmental quality as U.S. citizens choose, suggests requiring Mexicans to buy the same quality food, clothing and shelter as their northern neighbors but on a lower budget. The big question is whether environmental policies reflect the desires of the people directly involved or whether those policies are simply a way of subsidizing production at the expense of the environment.

Care must be taken in negotiating and implementing a trade agreement that trading partners do not use environmental regulation as an excuse to establish protectionist policies. On the other hand, trade agreements must recognize the possibility that environmental policy (or lack thereof) can be used as an implicit subsidy on production. If these caveats are recognized, policies that promote freer trade and environmental protection can complement each other and contribute to public welfare.

Tackling Environmental Problems Related to Trade

From a U.S. perspective, progress has been made in dealing with environmental problems since the beginning of the NAFTA negotiations. This has been accomplished through so-called "parallel" discussions and agreements, and by integrating trade-related environmental issues into the text of the agreement.

As promised by the Bush Administration, the first phase (1992–1994) of a binational border environmental cleanup program is now being implemented (U.S. Environmental Protection Agency 1992, 50). Mexico has committed to invest at least $460 million over the next three years in environmental projects along its northern border. About one-half of this funding is earmarked for sewage systems and waste water treatment plants in Mexican border cities (Embassy of Mexico 1992, 25). President Bush requested a budget of $240 million to protect the border area environment in 1993. While these funding levels are small relative to the overall needs of the region, the increase in cross-border cooperation resulting from the NAFTA process is noteworthy. Building on the 1983 La Paz Agreement, Mexico and the United States are now working together to achieve an important common goal: the long-term protection of human health and natural ecosystems in the border region. The NAFTA negotiations have also led to more cooperation on the environment between Canada and Mexico. A major boost to this cooperation occurred in March 1992, when Canada's Secretary of State for External Affairs and Minister of the Environment announced projects valued at $1 million to assist Mexico in environmental monitoring and enforcement (NAFTA Environmental Review Committee 1992, 94). The Canada–Mexico environmental initiative will strengthen Mexico's capability to enforce its environmental law. It will also demonstrate Canadian public and private sector expertise in environmental technology, thereby opening the door to future commercial collaboration between the two countries.

For its part, Mexico is striving to improve its environment. In 1988, Mexico enacted a comprehensive environmental law called the General Law on Ecological Equilibrium and Environmental Protection (General Ecology Law). This law, which broadly seeks to prevent and control pollution, is indeed similar to U.S. statutes. In some instances, such as requiring environmental impact statements prior to private construction, Mexico's law goes beyond U.S. requirements.

President Salinas has vowed that Mexico will not become a pollution haven for dirty companies from the United States or elsewhere. One indication of Salinas' commitment to the environment is that he has increased the federal government's environmental spending more than tenfold since 1989 (Embassy of Mexico 1992, 16) The Salinas government has cracked down on some industrial polluters. In 1991, the government permanently closed a major Mexico City oil refinery as part of a larger effort to improve air quality; and efforts are being made to direct new business investment away from heavily polluted metropolitan areas. President Salinas was awarded the 1991 Earth Prize by the Nobel family and the United Nations for outstanding environmental statesmanship.

In May 1992, the Mexican Congress approved the creation of the Secretariat of Social Development (SEDESOL). This new agency, headed by Luis Donaldo Colosio, is designed to integrate environmental protection with social and development policies. SEDESOL's environmental functions are divided between the National Institute of Ecology and the Office of the Attorney General for Protection of the Environment.

Under the direction of Dr. Sergio Reyes Lujan, the National Institute conducts research, supervises planning, and develops environmental regulations. Mexico's new attorney general for the environment, Santiago Onate Laborde, is responsible for investigating complaints and for enforcement. Top priorities include working with state and local governments to get more environmental inspectors on the job and increasing citizen participation in solving problems (Solis 1992).

Nevertheless, much can be done to improve enforcement of environmental laws in Mexico. For example, a recent U.S. General Accounting Office report (1992, 3) found that none of six new U.S. majority-owned maquiladoras that established plants in Mexico between May 1990 and July 1991 prepared environmental impact assessments as required by the Mexican government. To achieve better environmental protection, Mexico needs to devote additional resources to implement its General Ecology Law.

A review of the NAFTA text reveals that many of the 22 chapters contain provisions that could affect the environment. However, the specific commitments of the countries to address trade-related environmental considerations in the agreement are reflected in several key provisions. These environmentally-friendly provisions are presented in table below and discussed in the following paragraphs.

In the Preamble, the three countries make a commitment to "promote sustainable development" and to "strengthen the development and enforcement of environmental laws and regulations." Furthermore, the Preamble explicitly requires that all commercial objectives of the agreement be undertaken "in a manner consistent with environmental protection and conservation."

Each trading partner has signed several international environmental agreements, that contain specific obligations regarding trade and environment (e.g., the 1987 Montreal Protocol on Substances that Deplete the Ozone Layer). In Chapter 1 (Objectives), Article 104 ensures that the obligations of certain international environmental and conservation agreements to which the parties are members will take precedence over their NAFTA obligations. The international environmental and conservation agreements are: (a) the 1973 Convention on International Trade in Endangered Species of Wild Fauna and Flora; (b) the Montreal Protocol; (c) the 1989 Basel Convention on the Control of Transboundary Movements of Hazardous Wastes and their Disposal; (d) the Canada–U.S. agreement concerning hazardous wastes; (e) the Mexico–U.S. border area environment agreement; and (f) any subsequent international agreement that the parties agree shall be included.

In addition, should a disagreement arise concerning the implementation of Article 104, the responding party could elect to have the dispute settled exclusively under provisions of the NAFTA. The guarantees provided in Article 104 are a major exception to existing international trade law. For the first time, trade provisions in at least the named international environmental and conservation agreements would take precedence over the disciplines contained in an international trade agreement. There is, however, uncertainty concerning international environmental agreements that are omitted from the NAFTA text, and how future agreements will be handled by the parties.

The question of whether the NAFTA permits challenges to U.S. environmental laws or regulations requires review of its standards provisions. These provisions are found in Chapter 7, which sets out sanitary and phytosanitary measures, and in Chapter 9, which sets out all other standards-related measures.

Key Environmental Provisions
of the North American Free Trade Agreement

- The Preamble to the NAFTA specifically identifies environmental protection and conservation as a primary objective. It also identifies the promotion of sustainable development, and strengthening the development and enforcement of environmental laws and regulations.

- Chapter 1 (Objectives) includes a broad exception for specific trade obligations set out in certain international environmental and conservation agreements.

- Chapter 7 (Subchapter B: Sanitary and Phytosanitary Measures) would permit a NAFTA country to establish the level of protection that it considers appropriate to protect human, animal or plant life or health within its respective territories. Each party agrees to base its sanitary and phytosanitary measures on scientific principles and risk assessment, when appropriate to the circumstances.

- Chapter 9 (Standards-Related Measures) would protect the right of a NAFTA country to determine the level of environmental protection that it considers appropriate. It would require the three countries to work jointly on enhancing the level of environmental protection, and prohibits downward harmonization. The chapter sets forth broad legitimate objectives, and requires the creation of a committee to follow up on issues such as the development and enforcement of standards-related measures.

- Chapter 11 (Investment) contains an important provision that would formally discourage a government from lowering its own environmental standards for the purpose of encouraging investment.

- Chapter 20 (Institutional Agreements and Dispute Settlement Procedures) would provide new mechanisms for the submission of environmental concerns to dispute settlement panels, and would require the panels to take such concerns into account in making their decisions. It would also allow a responding party to require that any disagreement pertaining to a named international environmental or conservation agreement, or any standards-related measure affecting its environment, be considered exclusively under the NAFTA dispute settlement mechanism.

Source: Office of the U.S. Trade Representative (1992) and NAFTA Environmental Review Committee (1992, 10–36).

While there are major differences between the two chapters, the following language emphasizes that the NAFTA will not create new pressure for "downward harmonization," or otherwise cause a party to lose control of its own environmental and public health standards. For example, Article 754 states that:

- "Each party may . . . adopt, maintain, or apply any sanitary or phytosanitary measure necessary for the protection of human, animal, or plant life or health in its territory, including a measure *more stringent* than an international standard, guideline or recommendation . . . ."

Articles 904 and 905 specify that:

- Parties may adopt, maintain, and apply standards-related measures, including those relating to safety, the protection of human, animal and plant life and health, the environment and consumers, and measures to ensure their enforcement or implementation; and

- This provision shall not be construed to prevent a party in pursuing its legitimate objectives, from adopting and applying any standard-related measure that results in a *higher level* of protection than would be achieved through an international standard; and

- "Legitimate objectives" are defined to include safety; the protection of human, animal and plant life and the environment; and sustainable development.

Collectively, the provisions of Chapters 7 and 9 are an improvement over comparable provisions pending in the GATT negotiations. They strive for equality and fairness in trade, and restrain the ability of governments to use environmental regulations for primarily protectionist purposes. However, they do not prevent jurisdictions at any level of government from adopting and enforcing the level of environmental protection that they would deem appropriate. In Article 906, the text explicitly recognizes the role of standards-related measures in promoting legitimate objectives (defined above), and directs the parties to work jointly to enhance the level of environmental protection.

If a sanitary or phytosanitary measure is challenged as a trade barrier, the responding party does not have the benefit of the broad definition of "legitimate objectives" found in Chapter 9. Instead, the party must demonstrate that the measure is based on scientific principles and is no more restrictive than is necessary to maintain the party's "appropriate level of protection, taking into account technical and economic feasibility."

The NAFTA permits a party to prohibit the importation of any product that would harm its environment, or the health of its people, animals or plants, provided the same standards are applied to like domestic products and are not undertaken primarily to provide a commercial advantage to domestic producers. However, because the NAFTA seeks to follow GATT in rejecting the "extraterritorial" application of national law, it may be argued that a party is not allowed to enforce its process-related environmental standards in another country. Instead, the parties are directed to resolve problems dealing with process-related environmental standards on a cooperative basis through such mechanisms as the Committee on Standards-Related Measures. On the other hand, U.S. regulations designed to protect the environment fit within the legitimate objective exception to the "unnecessary obstacles" provision of Chapter 9. There are no provisions limiting the territoriality of this exception.

Chapter 11 (Investment) specifies that the NAFTA countries should not lower health, safety or environmental standards as a means of attracting investment. Article 1114 states that "the parties recognize that it is inappropriate to encourage investment by relaxing domestic health, safety or environmental measures." Furthermore, if a party has offered such encouragement, another party may request consultation. The two parties would then consult with a view of avoiding such encouragement.

This important provision is aimed directly at preventing the creation of "pollution havens" and avoiding subsequent pressure to lower environmental standards in other NAFTA countries. If successfully implemented, this provision means that environmental measures would take precedence over the investment objectives of the agreement under certain circumstances. A weakness is that the agreement does not provide an enforcement mechanism to ensure that the parties are complying with this provision.

Chapter 20 (Institutional Arrangements and Dispute Settlement Procedures) recognizes the importance of taking environmental considerations into account in the settlement of trade disputes. It provides new mechanisms—including technical expert and scientific review boards—for introducing environmental sciences before a dispute settlement panel. It places the burden of proof on the party that challenges the consistency of an environmental measure with the provisions of the agreement. If there is any doubt remaining with the dispute panel, the environmental regulation or policy wins out. Finally, Chapter 20 would allow a responding party to require that any disagreement pertaining to a named international environmental or conservation agreement, or any standards-related measure affecting its environment be considered exclusively under the NAFTA dispute settlement mechanism rather than under the narrower provisions of the GATT.

As in other trade agreements, the NAFTA's dispute settlement process is designed to resolve disagreements between governments. It does not allow citizens or environmental groups to submit unsolicited briefs, nor does it provide rules setting specific standards of review for the dispute process.

In addition to the above environmental provisions, the NAFTA broadens the environmental exceptions in the GATT to include environmental measures necessary to protect human, animal and plant life and health, and measures related to the protection of living and non-living exhaustible natural resources. The NAFTA text also includes extensive notification and publication provisions that would permit interested citizens and environmental organizations to influence the environmental standards of all parties to the agreement. Restrictions on technical experts working in the environmental sectors of the three countries are eliminated and all tariffs on environmental equipment would be removed within 10 years.

Consistent with the objective of maintaining sovereignty, the NAFTA recognizes the unilateral right of a party to protect its own environment. However, it does not provide a unilateral right to dictate the conditions of production or consumption in a foreign country. For example, the agreement does not give the United States government the right to enforce the same environmental regulations on U.S. firms operating in Mexico and Canada that exist in the United States. Nor, does it give the United States, or any trading partner, the right to levy countervailing duties on imports equal to the difference in the cost of complying with environmental regulations between trading partners. Lax enforcement practices of a trading partner are not recognized as explicit trade distorting subsidies.

Completing the Task

On December 17, 1992, Presidents Bush and Salinas and Prime Minister Mulroney each signed the North American Free Trade Agreement. If approved by the legislatures of the three countries, the agreement will enter into force on January 1, 1994.

In the United States, the fate of the NAFTA will be determined by the Clinton Administration and a new Congress. While there is no limit on the amount of time that may be used to prepare an implementing bill for the agreement, when the bill and its supporting documents are submitted to the Congress, the President will have the benefit of the "fast track" procedure. By statute, the Congress will have 90 session days to vote on the implementing bill with no amendments.

In an October campaign speech on international trade, then-Governor Bill Clinton announced his support of the NAFTA, but insisted that he would modify the implementing legislation and negotiate supplemental agreements with Mexico and Canada on worker's rights, health standards and environmental protection (Clinton 1992). Since his successful election, Clinton has met with President Salinas and agreed not to seek to renegotiate the text. In addition, he has selected Senator Lloyd Bentsen, a NAFTA proponent, to lead the Treasury Department.

The new U.S. Congress may be somewhat unpredictable on trade policy.

With Bentsen moving to the Executive Branch, the Senate Finance Committee will be chaired by Senator Daniel Patrick Moynihan of New York. In the past, Moynihan has opposed the NAFTA expressing reservations about negotiating "the first free trade agreement . . . with a country that isn't free" (Pastor 1993). In the House of Representatives, the Ways and Means Committee will have many new members. Some of these members probably campaigned against the NAFTA and others may wish to move slowly on the issue. The mood of the new Congress suggests that the Clinton Administration will allow ample time to deal with the concerns of key Committees.

Given this political setting, environmentalists will probably pursue one of two options. On the one hand, some environmentalists have already labelled the NAFTA text unacceptable and called for its renegotiation (*Washington Post* 1992). These people believe that the agreement will undermine U.S. environmental standards, and they are concerned that Mexico does not have ability to enforce its own environmental laws. It appears that they will urge the Congress to reject the NAFTA.

Other environmentalists view the NAFTA and its related parallel agreements as new commitments by the countries of North America to improve the environment. They will likely work with the Clinton Administration and the Congress to strengthen these commitments. For instance, in the United States, environmental concerns can be dealt with in a Side Letter Agreement between the parties, without reopening the NAFTA text; and in domestic implementing legislation.

The terms of the Side Letter Agreement should be resolved prior to the passage of implementing legislation. The agreement would include several priority issues. For example, the parties could extend the list of Article 104 international environmental agreements that take precedence over the NAFTA. The parties could also agree to provide automatic acceptance of future amendments to these agreements.

Another priority for the Side Letter Agreement is to define the relationship between the North American Commission on the Environment, created by the parties in September 1992, and the environmental provisions of the NAFTA. One possibility is to delegate to the commission the power to act as the NAFTA body with regard to Chapter 9 environmental standards and Chapter 11 investment issues. Such a delegation would increase the likelihood that the NAFTA's Committee on Standards-Related Measures will systematically improve enforcement programs and help resolve environmental issues relating to methods of production and processes. While the NAFTA rejects the "extraterritorial" application of national law, a party has a valid concern if weak process-related environmental standards in another country cause transborder or global damages. The parties should also agree on the specific steps the commission will follow in resolving these problems.

The Side Letter Agreement can be used to better define the consultation process called for in Chapter 11 when there is a concern that environmental standards are being reduced to attract investments. Furthermore, the consultation process might be extended to address the general problem of lax enforcement of environmental laws among the parties. For example, the trading partners would first commit to better enforcement of their own laws as in the Preamble. Then, they would agree to hear complaints about lax enforcement and to work together to resolve problems that distort business activity and harm the environment. A cooperative approach to better enforcement of environmental law is preferable to the use of trade sanctions or other measures that threaten the sovereignty of a nation.

To facilitate this extension of the environmental language in Chapter 11, the powers of the commission should be clearly delineated. With respect to both investments and enforcement, authority is needed to thoroughly investigate the allegations of a party and to prepare a formal recommendation. Also, each country should agree to contribute to an annual report and public hearing on environmental standards and enforcement practices across North America. Such a process would help the countries set priorities for solving environmental problems in a staged fashion over several years and would create an opportunity for citizen participation.

It seems likely that the U.S. implementing legislation will be drafted through consultations between the Clinton Administration and the Congress. This process may take several months. The implementing legislation can contain any provisions that the President and Congress decide to include in it. There may be some overlap with issues covered in the Side Letter Agreement.

A top priority of environmentalists will be to ensure that the implementing legislation obligates the U.S. government to fully fund its share of the United States–Mexico border plan and the North American Commission on the Environment. Funding might be provided through general appropriations and from earmarked revenue sources. For example, the U.S. Customs Service collects about $230 million per year from its value-added duty on imports from Mexico. Revenue from this duty could be dedicated to environmental projects for a period of time, say 10 years, and then eliminated under the terms of the agreement.

Dispute resolution and consultation under the NAFTA are initiated by governments of the parties, not by private individuals or organizations. Therefore, environmentalists are likely to seek a provision in the implementing legislation giving them the right to petition the U.S. government to challenge objectionable environmental practices of a party, and to become involved in formulating government policy with respect to a dispute. For example, allowing public participation in dispute settlement proceedings challenging U.S. environmental law is important.

Environmentalists will also seek provisions in implementing legislation: (1) stating explicitly that the agreement does not preempt federal or state environmental laws; (2) requiring U.S. investors operating abroad to fully comply with host countries environmental laws; and (3) directing U.S. Customs to block the import of products that do not meet U.S. health, sanitation and environmental standards. These provisions are consistent with NAFTA's goal that commercial activities of the parties be undertaken in a manner consistent with environmental protection and conservation.

Conclusion

President Salinas's decision to break with the past and seek to dismantle trade and investment barriers is a hopeful sign for Mexico's future. The principal problems on the environmental agenda in North America stem not from huge differences in policy, but from a gap in capabilities to enforce laws and from a need for better cooperation on transborder and global problems. Implementing an environmentally-friendly NAFTA and its related parallel agreements on the environment provides additional assurance that we can solve these problems.

Notes

1. From the outset the countries agreed to focus on trade in goods, services, intellectual property and investment; and to produce an agreement consistent with the GATT requirements for free trade areas.
2. It should be noted that second-order effects could be positive as well as negative. As development occurs, individuals not directly effected may derive benefits from knowing that resources are being put toward improving incomes in less developed countries. Full accounting would have to consider both beneficial and damaging second-order effects, neither of which is easily quantified.

References

Clinton, Bill (Governor). 1992. Remarks on expanding trade and creating American jobs. Raleigh, NC: North Carolina State University. (October 4): 1–19.

Collinge, Robert A., and Peter M. Emerson. 1992. The North American Free Trade Agreement: Short term obstacles and long term opportunities. *Operations Management Review* 9(1): 12–17.

Embassy of Mexico. 1992. *Mexico environmental issues.* Washington, DC, September.

Emerson, Peter M. 1991. Testimony before the Subcommittee on Commerce, Consumer Protection and Competitiveness of the Committee on Energy and Commerce. Washington, DC: U.S. House of Representatives. (May 8): 1–4.

Goebel, J. Martin. 1992. Clearing the air on Mexico's environmental track record. *The Wall Street Journal*, June 12.

Hair, Jay. 1991. Nature can live with free trade. *The New York Times*, May 19.

Kleist, Trina. 1992. Mexico's toxic legacy: Country fighting tide of dangerous wastes. *San Antonio Light* (November 1): D1, D2.

NAFTA Environmental Review Committee. 1992. *North American Free Trade Agreement–Canadian Environmental Review*. Ottawa, October.

Office of U.S. Trade Representative. 1992. *North American Free Trade Agreement*. Washington, DC: USTR, September.

Pastor, Robert A. 1993. NAFTA as the center of an integration process: The non-trade issues. *The Brookings Review* (Winter): 43.

Pearson, Charles S. 1987. Environmental standards, industrial relocation and pollution havens. In *Multinational corporations, environment and the Third World*, edited by Charles S. Pearson. Washington, DC: World Resources Institute.

Reinstein, R. A. 1992. Trade and environment. State Department, Washington, DC.

Solis, Diana. 1992. Environment officer has toughest job in Mexico. *The Wall Street Journal* (August 17): A4.

U.S. Congress. Office of Technology Assessment. 1992. *Trade and environment: Conflicts and opportunities*. OTA-BP-ITE-94. Washington, DC: U.S. Government Printing Office, May.

U.S. Environmental Protection Agency. 1992. *Integrated environmental plan for the Mexican–United States border area (first stage, 1992–1994)*. Washington, DC: EPA, February.

U.S. General Accounting Office. 1992. *Assessment of Mexico's environmental controls for new companies*. GAO/GGD-92-113. Washington, DC, August.

Washington Post. 1992. SABOTAGE! of America's health, food safety and environmental laws. (December 14): A20.

5

Trade Liberalization, Agricultural Policy, and Wildlife: Reforming the Landscape

J. H. Patterson

Introduction

Global and regional trade liberalization in the 1990s will have a profound impact on the economies of both developed and developing nations. While the Uruguay Round of the General Agreement on Tariffs and Trade (GATT) has bogged down with negotiations extending several years beyond the original schedule, the North American Free Trade Agreement (NAFTA) negotiations have been successfully completed between Mexico, the United States, and Canada.

Though there generally has been widespread support for both the NAFTA and GATT, two areas have proved to be stumbling blocks in the negotiations. First, since negotiations for the NAFTA were opened in June 1991, the impacts of trade liberalization on the environment have received more and more attention. In Mexico, the United States, and Canada, certain environmental and labor interests have strongly opposed the agreement based on the speculation that there will be a massive migration of jobs to Mexico and other developing countries with low wage economies. The environmentalists predict that Mexico and other developing countries will provide pollution havens for American and Canadian businesses. They also fear that Canadian and U.S. pollution and safety standards will be lowered to allow Canadian and U.S. exporters to be more competitive. They judge the NAFTA to be inappropriate because it does not contain specific command-and-control standards and regulations for environmental and employment protection. Trade negotiators see environment as being the leading trade issue of the 1990s.

A second stumbling block, particularly with the GATT negotiations, has been agricultural subsidies, which the Organization for Economic Cooperation and Development (1991) estimates are in excess of $300 billion annually for its member countries. The negotiating position of the United States and Canada is to remove all export subsidies on agricultural products, but the European Community (EC) has resisted any reduction in the high level of public support provided to agriculture under their Common Agricultural Policy. When pressure is put on the EC to make changes, farm groups lobby their respective governments to resist any lowering or modification of government subsidies. The resistance to dismantling agricultural subsidies prevails in spite of the overwhelming evidence that agricultural subsidies have distorted international markets, depressed commodity prices to record low levels, caused severe environmental problems in Europe, and encouraged the farming of marginal lands and wetlands, severely impacting wildlife populations.

Given that environmental concerns have taken center stage in the negotiation of trade agreements, it would appear that reductions in agricultural subsidies would be an obvious place to improve trade and the environment. The remainder of this chapter considers the potential for enhancing wildlife habitat through trade liberalization. In particular, it is argued that reduced agricultural subsidies brought on by free trade agreements will reduce the incentive to cultivate marginal lands and thereby increase wildlife habitat.

Background

In the older settled regions of eastern North America, patterns of agricultural land use have tended to reach an equilibrium between market forces and land capability. While extensive areas were cleared for farming during settlement, those lands that were marginal for crop production were soon abandoned. Now, high-quality farmland is used intensively for sustainable crop and livestock production, whereas the marginal lands have tended to revert to native vegetation. The resulting landscape is a diverse mosaic of productive farmland and natural habitat providing benefits for wildlife and environmental quality.

The development of these patterns of agricultural land management took place long before government interventions in the market place. They are cited as an example of the fact that free market agriculture results in a diverse landscape that supports a stable and productive agricultural industry while at the same time natural habitat for the benefit of wildlife. These patterns of agricultural land management, which have evolved over many years, represent sustainable utilization of landscape resources. The report of the Federal-Provincial Agricultural Committee on Environmental Sustainability defined sustainable agriculture as "agri-food systems that are economically viable, and meet society's need for safe and nutritious food, while conserving or enhancing natural resources and the quality of the environment for future generations." In the prairie and bottomland hardwood regions of North America, there has been considerable expansion of grain and oilseed production in recent decades. Much of this extensification has been onto marginal or fragile lands and wetlands that provided critical habitat for wildlife. Anderson and Leal (1991) have described the impacts of federal farm policy on land management in the United States. Direct government subsidies are the main factor responsible for agricultural drainage of wetlands and expansion of cropping onto marginal and fragile soils.

The Global Playing Field

Since publication of *Our Common Future* (World Commission on Environment and Development 1987) by the Brundtland Commission, the notion of sustainable development has gained conceptual acceptance by most sectors in Canada. Sustainable development is seen to be a product of linking environmental and economic factors into day-to-day decision making. More recently, biodiversity conservation has emerged as a globally significant environmental issue. Because the concept is not well understood in many quarters, there is a high degree of uncertainty and apprehension over its implications. To some, the environmental component of sustainable development has been overtaken by biodiversity. At one extreme, biodiversity objectives are thought to be achieved only through wilderness protection. At the other extreme, biodiversity objectives are seen to adversely affect sustainable economic development of natural resources. Both extremes imply significant economic and perhaps social cost.

In the context of this discussion, biodiversity conservation means the maintenance and restoration of viable plant and animal populations and the physical environment at levels that sustain essential ecological processes. An integral component of this definition is the presence of viable human communities based on the environmentally and economically sustainable utilization of natural resources.

The most substantial and far reaching global discussions on the environment and on the economy were held in 1992. The focus for these talks has been the United Nations Conference on Environment and Development (UNCED) held in Rio and the Uruguay Round of GATT negotiations in Geneva.

Though Rio and Geneva are miles apart, the purpose of this discussion is to demonstrate that they are not poles apart. UNCED '92 sought to chart a new course of action for global cooperation to achieve sustainable development and biodiversity conservation. Negotiations on the Uruguay Round of GATT seek to chart a new course of action for global trade liberalization by breaking trade distorting subsidies and barriers to international commerce. Following the successful negotiation of a NAFTA, North America is in the unique position to provide international leadership in linking Rio and Geneva. There is an opportunity to develop cost-effective, market-driven approaches to strengthening the continent's agricultural industry and rural communities while enhancing environmental quality and biodiversity conservation.

A Canadian Case Study

A recently developed integrated computer model has demonstrated that the area of greatest biodiversity risk in Canada is in the developed agricultural landscape (Rubec, Turner, and Wiken 1992). On a regional basis, biodiversity risk is most severe in the southern prairie provinces, where at least 25 of Canada's endangered or threatened species occur. The prairie and parkland region provide critical breeding habitat for up to 50 percent of North American duck populations. The significant decline in prairie Mallard and Pintail populations since the mid-1970s illustrates how these duck species serve as a barometer of an environment that has been under increasing stress (Canadian Wildlife Service 1992). This stress has had equally significant impacts on both the economic and social viability of rural communities in the region. The North American Waterfowl Management Plan calls for expenditures of $1 billion over fifteen years to protect and enhance wetland and upland habitat and restore duck populations to the levels of the 1970s.

The presence of farming per se has not caused this environmental and economic dislocation; rather, ecological and economic integrity have been impacted by the substantial expansion of cultivated acreage beyond the sustainable land base onto marginal lands and wetlands—the area of critical importance to biodiversity. Of the approximately 54 million hectares of total farmland in the prairie provinces, only 32 million hectares are defined as high quality farmland by the Canadian Land Inventory (Ministry of Industry, Science and Technology 1992).

Expansion of cultivated acreage in the 1970s beyond the sustainable capability of the landscape was in response to strong commodity prices and increasing export markets. It was also fueled by agricultural policy and support programs that were based on acreage under cultivation and commodity production. The prolonged drought of the 1980s revealed that the present system of agricultural production was not sustainable on much of the marginal land. Government support programs did not provide adequate risk protection so new safety net programs were designed. Unfortunately, these new programs were again based on acreage, yield and commodity prices. International trade disputes have resulted in depressed market prices, so that safety net programs and deficiency payments have become a progressively larger portion of total farm income. The end result is that production is being driven more by government programs than by market demands. The negative impacts of these actions are the continuation of grain and oilseed production on marginal lands and the reduced likelihood of agricultural diversification. It is only in recent years that the environmental, economic, and social consequences of these practices have been realized.

It is noteworthy that as international commodity subsidy wars increased through the 1980s and as markets and commodity prices declined, the further expansion of cropping on marginal lands continued. It is only in the 1991 agricultural census that the amount of crop and fallow acreage has stabilized in the prairies.

Canada is well placed in the development of policy frameworks for the transformation to environmentally sustainable agriculture. The agriculture and conservation sectors have worked closely for a number of years and are in general agreement as to what needs to be done to restore the environmental and economic sustainability of the agricultural landscape. In very general terms, there is agreement that agricultural policies and programs should be decoupled from commodity production to allow cropping decisions to reflect market signals and the sustainable capability of the land. There is also general agreement that marginal lands that have been converted to crop production should be retired to native vegetation and that incentives need to be provided for the adjustment to permanent cover and wildlife habitat.

Although policy and program forces are working against sustainable agriculture, there are encouraging signs that the acceptability of soil, water and wetland conservation programs by the farm community is real and growing. Federal and provincial soil and water conservation programs and wetland and upland conservation initiatives, such as the Prairie CARE program delivered by Ducks Unlimited under the North American Waterfowl Management Plan, are oversubscribed by farmers. In the past four years, over 1.3 million acres of marginal uplands and wetlands have been protected and restored to permanent cover. In spite of these gains in soil, water, and wetland conservation, overall land use decision making is overwhelmed by the unintended impacts of commodity based agricultural support programs and policies. Grain and oilseed production in the prairie provinces currently receives approximately $45 per acre per year in government support. However, through permanent cover programs and Prairie CARE, farmers have demonstrated, in the free market, a willingness to retire marginal lands for an annual incentive payment of $15 per acre. Because of much reduced input costs, farmers realize an increased net cash flow from the $15 per acre incentive to retire marginal lands. The potential savings to governments through expanded marginal land retirement programs are in the order of hundreds of millions of dollars per year. In addition, the environmental and wildlife benefits from marginal land retirement are estimated to be of similar magnitude.

There is a growing recognition that reform of Canada's agricultural policies and programs is essential if we are to move towards: (1) sustained and diversified agricultural production systems that are driven by market forces and land capability; (2) diversification of land based economic activities with environmental, recreational, social and cultural amenities to strengthen rural communities; and (3) maintenance and enhancement of environmental quality, wildlife habitat and biodiversity on agricultural landscapes. Unfortunately, the impacts of international agricultural commodity trade wars to Canadian agriculture are so distorting that any change is often perceived as a threat to the industry.

Opportunity

The NAFTA and the current text of the Uruguay Round of the General Agreement on Tariffs and Trade call for reductions in export subsidies of agricultural commodities and a realignment of domestic support programs so they do not distort trade or commodity production. The GATT text calls for the phased reduction in agricultural support payments over a six year period amounting to 20 percent and 36 percent for domestic and export support programs respectively. In Canada, these reductions would grow to approximately $1 billion annually by the sixth year. The requirement to reduce these subsidies, or realign them to meet "Green Box" criteria would provide a strong impetus to modify agricultural policy and support programs to encourage sustainable development. The phrase "Green Box" refers to those subsidies that are allowed under the GATT agreement, and include support for environmental measures, research, and rural infrastructure. None of these subsidies can be directed to increasing agricultural production.

Effective wildlife and biodiversity conservation in the agricultural working landscape can only come about through revitalized rural communities supported by more diverse and stable income opportunities that are economically and environmentally sustainable. Implementation of both the NAFTA and GATT may provide a unique opportunity for North America to revitalize the agricultural industry and work towards rural renewal.

There are three broad policy options for government support of prairie agriculture. Canada could theoretically move to reduce commodity support unilaterally. However, if the international agricultural trade wars continued, Canadian agriculture would be totally devastated. An extensive abandonment of farmland would probably have a net benefit to the environment, but the social and economic costs would be so staggering that they would be unacceptable to society.

In the absence of an agreement to the Uruguay Round of GATT, a second option would be for the United States, the European Community, and Canada to maintain the status quo by continuing current levels of domestic and export subsidies. This option would extend pressures to exploit marginal lands and wetlands and have severe impacts on landscape productivity. Both short and long term social and economic costs would be substantial.

The third option would be to decouple farm support programs from commodity production, and realign soil, water and wildlife conservation funding and policies with liberalized market forces. Such actions together with successful trade liberalization under the NAFTA and GATT could lead to public policies and international markets working together as a sustainable development market force. By removing the trade and land use distorting effects of commodity subsidies, agricultural land management decisions could move to a non-distorted equilibrium between free markets and land capability.

Summary

Potential elements of trade-driven adjustment to agricultural policies and programs that may contribute to sustainable agriculture, rural renewal, and biodiversity conservation are as follows:

1. Market forces: As international agricultural production subsidies are reduced or reprofiled, it is anticipated that commodity prices will increase. This should encourage land use decisions that are more responsive to market forces and to the sustainable capability of the land base.
2. Non-distorting commercial income support: Agricultural safety net policies and programs can be modified to remove trade and land use distortions and comply with international trade agreements. Decoupling support from commodity production to broader farm income should encourage a shift from gross production to sustainable productivity.
3. Conservation incentives: A portion of the $1 billion trade war peace dividend in Canada could be used as financial incentives, specifically for conservation measures, to help rural Canada and the agricultural industry adjust to environmentally sustainable and economically sound practices.

Positive actions in responding to trade liberalization could be a cost effective and affordable imperative for North America. It is possible that the agricultural industry, rural communities, and the environment all could be strengthened with the current expenditure levels. Inaction or maintenance of the status quo would inevitably contribute to an environmental, economic, and social liability of growing dimensions.

References

Anderson, Terry L., and Donald R. Leal. 1991. *Free market environmentalism*. San Francisco: Pacific Research Institute for Public Policy.
Canadian Wildlife Service. 1992. *Breeding ground waterfowl survey*. Ottawa.
Minister of Industry, Science and Technology. 1992. *Census overview of*

Canadian agriculture, 1971–1991. Ottawa.

Organization for Economic Co-operation and Development. 1991. *The state of the environment.* Paris.

Rubec, C. D. A., A. N. Turner, and E. B. Wiken. 1992. Integrated modelling for protected areas and biodiversity assessment in Canada. Paper presented at third national symposium of the Canadian Society for Landscape Ecology and Management on sustainable landscapes, June 17–19, Edmonton, Alberta, Canada.

The World Commission on Environment and Development. 1987. *Our common future.* New York: Oxford University Press.

6

Swapping Debts-for-Nature: Direct International Trade in Environmental Services

Robert T. Deacon and Paul Murphy*

Introduction

Discussions of the North American Free Trade Agreement (NAFTA) have focused on the potential for trading traditional goods and services and on possible negative impacts the NAFTA may have on the environment, but have ignored the possibility that the NAFTA might stimulate trade in environmental amenities and thereby foster environmental protection. The transaction costs associated with international trade in cars, vegetables, and steel pale in comparison to the costs of arranging direct trades in environmental amenities. In the latter case, one must deal with free rider problems, complications of specifying the goods and services involved, and issues of national sovereignty. The degree to which such problems can be overcome will depend, in part, on how innovative environmental entrepreneurs become in structuring contracts in environmental services. The NAFTA may well foster a political climate in which transaction cost barriers to trade of all sorts are lowered, including direct trades in environmental amenities. If so, then it will widen the scope for

* Thanks are due to Randall Curtis of The Nature Conservancy, Barbara Hoskinson of World Wildlife Fund, Ian Bowles of Conservation International, and Chris Herman of the U.S. Environmental Protection Agency for supplying information. We are indebted to Anthony Scott, Lee Alston, Alan Collins, David Simpson, and Ted Frech for helpful comments. The views and interpretations expressed here are our responsibility, however.

externally financed environmental protection measures in Mexico. Such arrangements are exemplified by debt-for-nature swaps, transactions in which conservation groups or government agencies in developed nations finance environmental protection in the developing world.

If transaction costs were negligible, there would be substantially more trade in the services of environmental assets. For example, countries of the Amazon basin might trade the provision of biodiversity and carbon dioxide absorption to the developed nations in return for higher education, electronics, and automobiles. The potential for gains from trade would be high because there are large differences in the environmental endowments of developed and developing countries, and because of differences in the commodities demanded by consumers in the two regions. Though there is nothing inherently suspect about trade in environmental services, the prospect seems strange because such trades are seldom observed.

Trades in environmental amenities are rare because high transaction costs prevent relevant markets from emerging. These high transaction costs stem partly from the fact that many amenities are subject to free riding and partly from the practical difficulty of policing the use of resources such as forests, water, and air. In the past these costs typically have outweighed the benefits of specifying contracts to exchange goods or services for environmental amenities, and the efficiencies that trade can provide have not been realized.

This situation is changing, however. On the benefit side, the value of establishing and exchanging rights to the developing world's environmental assets is increasing. Growing concern over increases in atmospheric carbon dioxide has raised the value of standing tropical forests, which remove carbon from the atmosphere and sequester it in their biomass.[1] Additionally, when forestlands are converted to commercial ranching or cultivated for agriculture, the standing biomass is often burned causing the immediate atmospheric release of a store of carbon that had accumulated over the ages. A second reason for increased trade in environmental amenities stems from recent advances in biochemistry that have enhanced the return from using the genetic information found in tropical forests for commercial purposes, such as new medicines, pest controls, and hybrid plants.[2] Thus, ethical objections to extinguishing species of plants, animals, and insects have been augmented by a potential profit motive for species preservation.

On the cost side, three factors can potentially increase the prospect for capturing gains from trade in environmental assets. First, free trade agreements such as the NAFTA raise the value of a good reputation to all potential traders. The risk that promises of environmental protection may not be kept after payments are made is one factor that deters trade in environmental services. The NAFTA will enhance the gains from future trade and thereby raise the value of maintaining a reputation for honoring trade contracts. For this reason, it may thus lower transaction costs for environmental trades. Second, an expanded volume of international trade may foster the development of third party mechanisms for enforcing international contracts. Third, the emergence of high altitude satellite imagery has provided a new, often inexpensive, source of information on how forests and other environmental assets are used. This, in turn, can allow owners to enforce claims to such resources at lower cost.

While these benefit and cost factors clearly expand the scope for trade in environmental services, they also enhance the incentive of current nominal owners to enforce their ownership claims and to prevent the wastes that can accompany free access. As one U.S. biologist with extensive field experience in conservation projects in Costa Rica put it, "if people say 'biodiversity has value' . . . then it will fall under the social rules that all other things that have value do. You bargain for it, you hide it, you steal it, you put it in the bank. It's no longer the toy of the English rich" (Joyce 1991, 39).

For these reasons ownership rights to environmental assets and services are being established and traded for the first time. One manifestation of this is the debt-for-nature swap, a contract between two or more parties to provide protection for environmental assets or to enhance the provision of environmental services. Contractual agreements concerning the identification and commercial application of biological substances found in the tropics are another manifestation. Other important examples are international treaties between nations for protection of the global environment.

In what follows, the transaction costs involved in trading environmental services are examined. The discussion draws cn experiences gained primarily in Latin America and, to a lesser extent, in Africa and Asia. The central theme in this discussion is that environmental trade contracts tend to be structured to minimize transaction costs. The parties to debt-for-nature swaps and other environmental exchanges gain by formulating contracts in ways that minimize the expected costs of monitoring and enforcing the contract's terms, the cost of resolving any disputes that may arise, and the costs of bearing any risk associated with the assets in question. The potential for evolving property rights and trade in environmental assets is illustrated in the next section with examples involving the commercial use of biodiversity. The third section outlines the history of debt-for-nature swaps, and traces their roots to the developing country debt crisis of the 1980s. This is followed by a discussion of the structure of debt-for-nature agreements completed to date, using the transaction cost minimization principle as an organizing theme. The final section concludes and argues that the NAFTA, by helping to reduce the transaction costs that inhibit trade in environmental amenities, may play a positive direct role in the protection of environmental assets.

Transaction Costs and the Evolution of Ownership

The transaction costs that hinder trade in environmental services arise from several sources. One source is the difficulty of obtaining payment from all who benefit from provision of an environmental service. For example, if a single country acts to mitigate ozone depletion, others will benefit regardless of whether they pay a share of the cost because they cannot be excluded from consuming the service once it is provided. This is the familiar free-rider explanation for under-provision of environmental services. A second source is the problem of exercising practical control over the use of environmental resources. If one seeks to enhance biodiversity and mitigate the greenhouse problem by protecting a tropical forest, it is necessary to observe and control how and by whom the forest is used. A typical forest's sheer size, remoteness, and multiplicity of access points makes such control costly, and similar problems apply to other environmental resources. The third reason that environmental transactions are more costly derives from their international dimension. To the extent that capital flows across international borders are restricted, negotiating any trade and making the payments agreed upon is more costly. Moreover, enforcement of contracts is complicated by the sovereignty issue and by the volatility of legal and political institutions of some less developed countries.

If trade in a particular good does not occur, the reason is more likely to be these transaction costs than the absence of legal title. Formally recorded ownership rights to a good tend to emerge only when the benefit that a single agent or cooperating group can reap from enforcing control rises above the associated cost. In this sense the existence and force of property rights is more a matter of economics than law. Nominal property rights to a resource are effective only if the benefit of enforcement to the party who seeks to exercise control outweighs the cost (see Barzel 1989, 65). This claim must be qualified in two ways, however.

First, it is predictive rather than normative. While rights tend to evolve when the benefits of ownership exceed the costs, this benefit-cost calculus is performed at the level of the individual and it may or may not be efficient in a broader social sense (see de Meza and Gould 1992). Second, because the entire process takes place within the matrix of legal and political institutions of a given country, those institutions may prevent socially efficient property rights from evolving.

Transaction Costs and the Structure of Contracts

The transaction costs that arise in a specific trade clearly depend on the nature of the good or service involved, but they also are affected by the way the transaction is structured. The choice of contract structure affects the costs of monitoring and enforcing the contract's provisions and the expected cost of settling any disputes that may arise after an agreement is struck. Because all such costs detract from the value of the transaction to the parties involved, the participants to an exchange can gain mutually by adopting a contract structure that mitigates these costs.[3] Examples of such mitigation include:

1. not specifying all contingencies because the cost of anticipating these and negotiating mechanisms for dealing with them may be high (see Barzel 1989, 68).
2. structuring a contract to lower the stakes each party has in any possible dispute and to constrain the latitude the parties have in attempting to gain advantage in the settlement process.[4]
3. shortening a contract's duration and making it renewable in order to forestall opportunism by placing at risk the mutual benefits of contract renewal and future exchanges.
4. specifying contracts in terms of the *inputs* to be used, their quantities, method of application, and so forth in order to reduce the costs of measuring and monitoring contract performance.
5. exploiting patterns of complementarity and substitution among goods so as to minimize the cost of achieving a particular goal. For example, if one wishes to reduce fuelwood gathering but finds it difficult to monitor directly the actions of individual gatherers, it may be cheaper to subsidize the price of a substitute fuel.[5]
6. agreeing that demanders of services will subsidize the enforcement of property rights to environmental assets owned by others. Hence, while a host government may nominally own a tropical forest and expend some effort to prevent deforestation, environmental groups or pharmaceutical companies may gain by helping to enforce the government's claim.

Exchanging Rights to Biodiversity

Recent agreements involving the commercial use of natural biological resources found in the tropics illustrate several of the preceding points. Although the volume of these trades has been small to date, a minor digression to examine them is worthwhile because it provides examples of practical approaches to the transaction cost problems encountered in contracting for environmental amenities.[6]

About one-fourth of all prescription drugs used in the United States are based on plant or microbial extracts or derivatives. Leading examples are quinine, isolated from the bark of the cinchona tree and used in combating malaria, and digitalis, obtained from the foxglove plant and used in heart treatments. More recently, the anticancer agent vincristine was discovered in the Madagascar periwinkle, the immunosuppressant cyclosporin was obtained from a Norwegian fungus, and invermectin for killing parasitic worms was isolated from a Japanese mold. Because tropical plants have survived by evolving chemical defenses against predators, they are a particularly rich source of useful substances. Only a small portion of their potential has been researched, however.

Technological change has reduced the cost of screening natural compounds for possible pharmaceutical uses at the same time that the demand for these natural compounds has increased. Hence, it is not surprising that efforts have been undertaken to establish property rights to them. An example is the contract between Merck and Company, a pharmaceutical producer, and the Instituto Nacional de Biodiversidad (INBio), a Costa Rican organization active in science and conservation, to conduct "chemical prospecting" in the parks and nature reserves of Costa Rica. Prior to the agreement, INBio had begun a 10-year project to catalog the country's estimated 500,000 species of plants, insects, and microorganisms. The contract specifies that Merck provides $1 million over two years to aid INBio's conservation efforts, e.g., for training parataxonomists to collect species and curators to catalog them and prepare extracts for testing. Merck receives the exclusive right, for two years, to analyze a specified number of samples from INBio's inventory, for possible commercial applications.

Negotiation of the Merck-INBio contract was costly, particularly due to the task of delineating the details of collecting and cataloging specimens. Merck's concerns are to structure a contract that guarantees the purity of specimens and accuracy in documentation of the season, time, and location of collection. The problem for INBio, as an agent for Costa Rica, is to structure a contract that encourages research by Merck but assures the country an attractive return if Merck develops a highly lucrative substance. The INBio-Merck agreement resembles a *wage contract* in the sense that Merck pays a fixed sum in return for collecting and screening services supplied by INBio. This implies that all of the variance in the project's returns, which may be substantial for chemical prospecting, rests with Merck.[7]

Other contract forms are also being tried. Under an agreement between INBio and two universities, Strathclyde in Scotland and Cornell in the United States, INBio tests indigenous compounds for pre-specified chemical characteristics and sends promising samples to Cornell or Strathclyde, where they undergo further analysis.[8] If Cornell or Strathclyde, or a commercial partner of either, develops a patentable substance from any of these, INBio can claim between 51 and 60 percent of the patent royalty, depending on the amount of modification needed to obtain the final product. An alternative contract form, termed "the lottery," is also under consideration. Under this option, INBio would send several hundred coded compounds to an interested pharmaceutical company. If the company finds a potential winner in the package and wants further evaluation, it must agree to share royalties with INBio before the source of the substance is revealed and additional supplies are provided. The lottery proposal and the agreement between INBio and Cornell and Strathclyde Universities are *share contracts* in the sense that they split the gross value of the activity's output between the parties. This exposes INBio to the risk that Cornell and Strathclyde may not be diligent in the research and legal work necessary to see a compound through the patenting process, because sharing dilutes their stake in the outcome. The share contract allows risk to be spread, however, and this increases the value of the resource if the two parties are risk averse. The share contract also encourages diligence in the collection effort, since INBio has a partial stake in the value of output.[9] The natural alternative to these contractual arrangements is for host countries to develop and market their own biodiversity directly, effectively integrating collection, cataloging, screening, and pharmaceutical development in a single organization. This option presumably would be chosen if the costs of organizing and overseeing these activities internally are deemed lower than the costs of transacting them in a market. Indeed, Joyce (1991, 36) reports that Costa Rica intends to develop the capability to screen natural compounds itself.

Swapping Debts-for-Nature: A Brief History

An alternative to commercial contracting to protect environmental assets in tropical forests are debt-for-nature swaps, of which 21 have been negotiated in 11 countries as of November 1991.[10] Those involved in these transactions cite the availability of a secondary market for "nonperforming debt" and discounted debt prices, as low as five cents on the dollar, as necessary ingredients for interest in these swaps. The secondary market arose from the debt crisis that became evident in 1982, when Mexico suspended its debt service payments. Several Less Developed Countries (LDCs) followed, rescheduling debt payments as worldwide recession and high interest rates made debt obligations increasingly burdensome. Consequently, total debt outstanding to developing countries increased from $551 billion in 1982 to $727.7 billion in 1985 and eventually reached $1.34 trillion by 1990.[11]

The secondary debt market arose under a more general program of "active debt management." Many private banks were heavily exposed in 1982 and willing to exchange debt in one form or another at large discounts. The volume of LDC debt traded between 1982 and 1987, around $10 billion, was small in comparison to the amount outstanding. Prior to 1987, interest in selling at a discount was limited both by accounting conventions and by a belief in the long-term potential of LDC debt.[12] Between 1984 and 1988, however, secondary market prices for Third World debt plunged, thus stimulating interest in swapping debt for nature.[13] The first swap, between Bolivia and Conservation International, was organized in 1987, three years after the suggestion by Dr. Thomas Lovejoy of tying debt reduction to nature preservation.[14]

Private Debt-for-Nature Swaps

Nineteen of the 21 debt-for-nature swaps transacted through November 1991 have used funds raised, at least in part, by private international conservation organizations (COs). With the exception of the first swap between Bolivia and Conservation International, each transaction has involved at least one CO in the host country, as well as host country government agencies.[15] The three international COs most active in debt-for-nature swaps are Conservation International (CI), The Nature Conservancy (TNC), and the World Wildlife Fund (WWF), although other organizations have been involved at one time or another in donating debt or obtaining debt with donated funds.[16] The debt reduction accomplished by these swaps totaled $99 million as of November 1991 and the cost of debt acquired amounted to $17 million.[17] Costa Rica and Ecuador have retired $80 million and $10 million, respectively, accounting for 90 percent of the total debt retired using debt-for-nature swaps. Overall, the total number of swaps per year increased from two each in 1987 and 1988, to five or six per year between 1989 and 1991. The volume of debt transactions peaked in 1989, with around $44 million in debt (face value) retired.[18] Starting in 1990, COs appeared to be initiating swaps with a greater variety of countries, but at smaller debt values per swap.

Public Debt-for-Nature Initiatives

Debt-for-nature swaps using funds raised by governments have involved Sweden, Holland, and the United States. The first such swap was funded by Holland in 1988 and involved $33 million face value of Costa Rican debt bought at a discounted price of $5 million. A swap between Sweden and Costa Rica in 1989 reduced Costa Rica's debt by $24.5 million at a cost to the Swedish government of $3.5 million.

Transactions involving U.S. government have proceeded under the Bush Administration's Enterprise for the Americas Initiative (EAI).[19] The EAI authorizes the President to reduce and restructure a country's P.L. 480 "Food for Peace" debt in exchange for economic and environmental concessions by Latin American and Caribbean countries.[20] Three EAI agreements have been completed to date involving Chile, Bolivia and Jamaica, and the face value of debt reduction equals $259 million.

The Global Environment Facility (GEF) was organized by the United Nations Development Programme, the United Nations Environment Program, and the World Bank in 1991. It is a three year pilot program of grants and loans designed to help developing countries bear the cost of global environmental protection. Funding under the GEF is granted only for projects that benefit the global environment, as distinct from the local environment, and are targeted on four problems: global warming, biodiversity, pollution of international waters, and atmospheric ozone.

The Conversion Process

The following outline describes a typical debt-for-nature swap, one financed by funds raised by an international CO. The host country participants usually include a local CO with whom the international CO has established a working relationship, a local government body with responsibilities and oversight that vary from country to country, and the central bank which must be willing to convert external debt to domestic currency obligations.[21] The process begins when a debtor country first approaches the CO or when a debtor gives approval to a CO to negotiate a swap. The international CO and debtor government negotiate the exchange rate to be used in debt conversion, the management of the conservation program, and the plan for expenditures of funds. The actual debt instrument is acquired only after an agreement is reached. Acquisition is sometimes complicated by covenants entered into during debt rescheduling, provisions that prevent a bank from selling the debt of a particular country without the permission of other banks that hold its debt. The CO can either buy debt using funds from donations, or receive the debt directly from a bank as a charitable contribution.[22] The next step is to transfer title to the debt note and to accomplish the conversion per the agreement. The exact method of transfer may depend on tax considerations. The debt may either be converted into interest paying bonds, with the interest used by the local conservation group for qualifying projects, or exchanged for promises of government conservation actions. Lastly, the swap must be executed, i.e., the agreed-to conservation actions must be carried out. Enforcing these contracts can be difficult because they involve sovereign governments, and there is no international body with authority to mediate and enforce should a discrepancy arise in executing the agreement.

Practical Significance of Debt-for-Nature Swaps

The amount of debt relief provided by debt-for-nature swaps to date has been small. At the time of the first swap there was $300 billion of tradable Latin American debt, and total commercial bank debt to the world's 15 most indebted countries was at $430 billion. To date, debt-for-nature swaps have eliminated only $99 million.[23] While it is highly unlikely that debt-for-nature swaps will make a major dent in Third World debt levels, it appears that they can make an important contribution to conservation and resource management in the countries involved. In many cases swaps finance projects that would not otherwise receive any funding. The first swap Ecuador undertook with WWF in 1987 reduced Ecuador's $8.3 billion foreign debt by only $1 million. Yet it established a conservation fund that provided "annual financing twice that of the existing government park budget" (Patterson 1990, 6). Alvaro Umana, Costa Rica's Minister of Natural Resources, Energy and Mines, pointed out that "The interest alone from Costa Rican debt-for-nature swaps is several times more than the annual budget of our national park service" (Reilly 1990, 136).

Aside from the potential benefits of debt reduction and enhanced biodiversity taken separately, many conservationists believe there is an added benefit from joining *debt* and *nature* in the same transaction. It is frequently claimed that debt causes deforestation and other forms of environmental degradation. Conservationists argue that developing countries are forced to exploit their natural resources in an attempt to service foreign debt and hence claim that there is a direct link between debt and the environment. The temptation to attribute a causal relationship between high levels of debt, rising agricultural exports, and deforestation is obviously strong for those who wish to promote debt-for-nature swaps, and the conjecture may in fact be true. To date no compelling evidence of a special synergy from including debt relief and promises of environmental protection in the same transaction has been presented, however. An alleged debt-environment connection is now being used to argue for Third World debt relief on environmental grounds, as a way of preventing the destruction of environmental resources that creditor nations seem anxious to preserve. At a very minimum, this relationship needs to be established and rigorously documented before such claims are translated into policy actions.[24]

The Structure of a Debt-for-Nature Agreement

Regardless of the dollar sums involved or the intentions of the participants, the actual accomplishment of environmental objectives requires that the transaction costs that universally hinder efficient use of environmental resources be overcome. Mitigating these costs requires that debt-for-nature agreements be structured with an eye to the kinds of monitoring and enforcement costs that such transactions involve.

The hypothesis that contract structures evolve in ways that mitigate monitoring, enforcement, and dispute resolution costs is used as an organizing theme in what follows. The nature of these costs and hence the contract form that best mitigates them depends on the enforcement options of the parties involved. Although there are variations, certain features have become standard in the debt-for-nature swaps negotiated to date so it is sensible to examine a single representative document. An agreement involving the World Wildlife Fund, Inc. (WWF) and Costa Rica is used as an example in what follows.

A Representative Debt-for-Nature Contract

On March 20, 1990, a debt-for-nature swap was concluded between WWF, the Costa Rican Ministerio de Recursos Naturales, Energia Y Minas (the Ministry), and Fundacion de Parques Nacionales (the Foundation).[25] The Ministry is a resource management agency of the Costa Rican government and the Foundation is a private, non-profit conservation organization in Costa Rica. Prior to completion of the swap, the Central Bank of Costa Rica announced its willingness to accept exchanges of Costa Rican external debt in amounts up to $10.8 million in trade for Costa Rican government bonds denominated in domestic currency, with the proviso that the bond proceeds be used for domestic conservation activities.

The contract opens with a number of preliminary statements, including references to the country's unique natural resources, the Central Bank's intent to fund conservation projects, the conservation objectives of the parties, and a separate trust agreement that establishes a conservation fund and names a private bank as trustee.[26] Next, the contract charges WWF to acquire Costa Rican debt instruments up to a specified limit and directs the Bank to exchange them for Costa Rican government bonds, to be held in the trust account just mentioned.[27] The prescribed uses of funds are then delineated in broad terms: "planning, administration, protection, and management of protected areas and their buffer zones," with more specific examples such as boundary demarcation, elaboration of management plans, development of infrastructure, and other activities related to nature interpretation or environmental education. The contract also specifically allows "training of a cadre of conservation professionals . . . to improve the local capacity for protecting and managing Costa Rica's natural resources." With these general goals established, the contract grants discretion to the two primary participants, WWF and the Foundation, to select, administer, and monitor specific projects. Project proposals are to be submitted by the Foundation and approval is dependent on WWF consent. The Foundation also is responsible for preparing budgets and reporting on activities completed.[28] The document then ends with assurances that projects approved will be compatible with the policies of the national government.

Swapping Debt for Enforcement of Ownership

To a large degree the conservation funds spent on debt-for-nature swaps are used to buy delineation and enforcement of nominal property rights that already exist and are held by others. The governments of many developing countries have set aside land in legal reserves, with title held by government, but failed to provide monitoring and enforcement. These parks exist on paper, but the governments who are nominal owners often find that the costs of enforcing their claims outweigh the benefits. As a consequence, they are effectively open access resources. The government's inability or unwillingness to control use is understandable. In many of the countries involved the populace is poor and ranks environmental protection a low priority, since preserving lands means giving it up as a source of food, fuel, or shelter. One Colombian summarized this point aptly by describing her country as one "where people literally eat biodiversity so as not to starve" (Sanchez 1992, 55).

Support for the central contention that these swaps are largely concerned with defining and enforcing pre-existing nominal property rights is found in the prescribed uses of conservation funds in debt-for-nature contracts. Marking boundaries and establishing park buffer zones are clearly enforcement actions.[29] Formulating a management plan delineates which uses of a resource are allowed, and hence constitutes a definition of rights. Other swap contracts specify spending funds to train and equip personnel to reduce illegal logging, wildlife poaching, and habitation. What is most unusual about these actions is that the international and host country COs are responsible for providing this delineation and enforcement, but neither party holds nominal title to the assets. Rather, legal title normally rests with the host government. In general, enforced ownership by *someone*, even if not the COs who donate funds, results in an environmental outcome that the COs prefer to open access.

Delineating Inputs Rather than Outputs

Because the international CO that raises funds seeks to preserve specific environments and prevent their degradation, one might expect a debt-for-nature contract to spell out in detail the degree of environmental protection to be attained. Using this structure would be problematic, however, because environmental quality is a complex, multi-dimensional set of attributes. Defining degrees of preservation during contract negotiation would be difficult and the process of monitoring and enforcing compliance after the contract is signed would be costly. Likewise, the difficulty of judging compliance unambiguously would raise the likelihood of costly disputes.

Transaction costs can be reduced by choosing a contract structure that delineates *activities to be undertaken* and *inputs to be applied*, rather than environmental outcomes to be achieved. The typical debt-for-nature contract describes a fund to be endowed and the allowed uses of it. Specific project proposals are then described in terms of the amounts and uses of inputs to be supplied, e.g., provision of park guards and marking of boundaries.

Contract Structure and the Issue of Sovereignty

When a swap between Bolivia and CI was signed, the press incorrectly reported that CI had gained ownership of part of Bolivia's forests. Many Bolivians were outraged, and one government official involved in the deal explained to the U.S. press by asking: "How would you like it if the Japanese used your trade deficit to buy the Grand Canyon" (quoted in Hamlin 1989, 1082). The important economic message in such remarks is that negative popular reaction to foreign ownership, particularly if the asset is a national park or reserve, would make it politically very costly for a government to enforce a foreigner's legal claim. Since such claims would be insecure, the reward for any effort spent negotiating ownership of environmental resources would be small. Evidently, the parties involved have concluded that the costs of negotiating such terms outweigh the benefits, since all swap contracts avoid foreign ownership of land or resources, and most rely on host country organizations to implement its terms.

The issue of foreign ownership also pertains to the conservation fund a debt-for-nature swap creates. Once the country's debt has been canceled, the CO must be concerned that its government might seize the conservation fund, fail to service the bonds the fund holds, or reduce the fund's value by inflating the domestic currency. The inflation problem is easily handled by indexing interest payments or by specifying the size of the fund and all interest payments in another currency, e.g., U.S. dollars, while allowing payments to be made in the host country's currency at the market exchange rate. The chance that the bonds might not be honored remains, however. The probability of repayment is partly related to a country's financial solvency, but political factors are present as well. Popular opinion in debtor nations sometimes regards the debts accrued by past political regimes as illegitimate. Indeed, some have objected to debt-for-nature swaps because they lend an air of legitimacy to the actions of former leaders (UNESCO 1991, 7).[30] Clearly, the *expected loss* due to non-repayment can be incorporated in the terms of the swap, by adjusting the size of the fund created for canceling a given amount of debt. The chance of default also introduces an element of risk, however, and this is costly for the CO to bear.

Conservation groups can take certain non-contractual steps to mitigate these risks. They can diversify, by undertaking numerous small swaps in several countries and by spreading any large swaps carried out among several donor organizations. They can also avoid swaps in the most politically unstable countries. Although most of the money spent in swaps has involved deals with Costa Rica, risks have been spread because seven different conservation groups have donated funds. Also, casual observation suggests that swaps have been avoided in the politically most volatile countries.

Sovereignty and the general nature of international law have implications for debt-for-nature contracts that go beyond the strict issue of foreign ownership. International law lacks concrete sanctions that demand compliance. While international courts will rule to uphold clear contractual obligations, none of these tribunals has enforcement powers and the conservation groups who are likely litigants lack standing.[31] Absent credible third party enforcement, it is sensible to avoid contract provisions that might, under plausible circumstances, require a sovereign state to act in ways that are inconsistent with the self-interest of its leaders. Overall, the contract structure that has evolved requires only that the host government not disrupt the programs agreed to by the principal parties, the international and local COs. In turn, the likelihood of interference is minimized by avoiding conservation actions that clash with the government's self-interest.

The Absence of Enforcement Mechanisms

Because debt-for-nature transactions are experimental and the assets involved are hard to define, the possibility that a contract will not be honored to the satisfaction of all parties is substantial. Accordingly, one might expect the inclusion of such safeguards as a conflict resolution clause, waiver of sovereign immunity, default remedy, and provisions for choice of law and forum. None of the agreements negotiated to date includes such conditions, however, and none even mentions default or the chance that an agreement might fail.[32] While this might seem a glaring omission, it can be explained by the kinds of disagreements that are possible and the cost effectiveness of the enforcement options available.

Consider first the kinds of disagreements that might arise. To oversimplify slightly, a debt-for-nature transaction includes three parties: the international CO that raised money to buy the debt, the debtor government that promises to service the bonds that endow the conservation fund, and the local CO charged with using the fund for conservation projects. While all three parties face risks, the most obvious are those the international CO faces if the other parties deviate from the agreement. The international CO must relinquish the host country's debt in return for the government's promise to make payments to a third party, the host country CO. The latter, in turn, must be trusted to promote projects that the international CO supports. If the local group's performance is lacking, or if the debtor country's government seizes the fund or fails to make interest payments, the international organization's reputation with its donors could be damaged and its status with taxing authorities could come under scrutiny. As Greener (1991, 163) points out, the lack of explicit enforcement provisions means that "once the international [conservation organizations] have relinquished the debt instrument, they are no longer legally significant parties."

Two alternatives are available to deal with the possibility that the debtor government will renege on its funding obligation. The first is to take legal recourse, and if the international CO intends this approach, then it is necessary to include default remedies and sovereignty waivers in the contract. The international group would have to rely on the debtor government and its courts to enforce these provisions, however, and there is little incentive for them to do so. An alternative strategy is for the international organization to document any default that may occur and to "complain publicly and attract attention" (Gibson and Curtis 1990, 342). To a large degree, international COs specialize in such actions and their effectiveness is self evident. Compared to the alternatives, it is a relatively cheap way for them to impose costs on any government that reneges, particularly if done in a way that influences the policies of multilateral development banks and foreign aid agencies. At the same time, it advertises the costs of reneging to governments involved in other swaps.

The international CO's other risk is that the local group might fail to perform, e.g., by mismanaging funds, undertaking inappropriate projects, or displaying incompetence. Again, there are two ways the international group might deal with such possibilities. The first is to bring suit, probably in the debtor country, and to rely on the debtor government to enforce their claim. The political costs of enforcing such a claim would likely be high to the debtor government. The environmental assets involved would be regarded as the patrimony of the debtor nation and the foreign plaintiff would have to argue that the local conservation group is managing them inappropriately. An alternative strategy for the international CO is to avoid any further swaps with the local group involved. Arguably, this is not very costly for the international group since there are numerous substitute uses of its donor's funds. The developing world abounds with environmental problems and the prospect of having to avoid future dealings with a single local conservation group in a single developing country would not be terribly damaging. For the local group that loses future business, however, such a loss could well be devastating. In a sense, the local group posts an implicit reputational bond when it agrees to perform the duties outlined in a debt-for-nature contract. The value of this bond is the present value of future resources the group expects to receive from the international group involved, and probably other international COs as well. Compared to the other resources a typical developing nation CO has at its disposal, such a loss is enormous.

The host country faces risks as well, but they are not of the sort that could be mitigated by common contract enforcement clauses. Given the structure of the transaction, the debtor government faces no risk from default and has no obvious incentive to favor inclusion of a sovereignty waiver. Rather, its risks arise from the chance that the international or domestic CO will take an action or promote a policy that damages the government's political relationship with its constituents. Typical debt-for-nature contracts include assurances that projects approved and actions taken will be consistent with the policies of the national government, however, thus providing the host government a lever to suspend operations funded by the swap if it finds them unacceptable. The inclusion of this escape clause, together with the fact that the party at risk in this case is a sovereign and has a sovereign's enforcement powers, obviates the inclusion of common contract enforcement mechanisms.

Risks to the host country CO appear minimal, and in any case are not of the variety that could be mitigated by including explicit enforcement mechanisms in the agreement. One might fear that a debt-for-nature contract would require the host country group to carry out projects it opposes. This is not the case, however, as typical agreements require host CO approval for all fund expenditures. Indeed, the typical contract grants the host conservation group operational control of the fund created. At the operational stage, the international CO's role is primarily one of oversight and project approval.[33]

Conclusions

The debt-for-nature swaps carried out in Latin America and elsewhere offer lessons that Canada, Mexico, and the United States will find useful as they consider the enabling legislation and parallel agreements that all parties agree are a necessary part of the final NAFTA. In particular, debt-for-nature swaps show that the environmental services that Latin American countries can provide are valued by those in higher income countries. If this value can be translated into an effective demand and implemented through environmental exchange contracts with Mexico to preserve its forests, deserts, or even its air and water, gains from trade clearly will result. The debt-for-nature swaps negotiated to date are structured to reduce transaction costs, but the costs that remain are still very high. The kinds of actions prescribed in debt-for-nature contracts have been examined, together with the ways these actions are delineated, the presence or absence of enforcement mechanisms, and so forth. In general, the structure of these contracts seems a rational response to (1) the monitoring and enforcement problems inherent in the assets transacted and (2) the costs and effectiveness of the enforcement measures available to the parties involved. For the most part, debt-for-nature contracts rely on delineation of inputs rather than outputs, reflecting the high costs of unambiguously measuring outputs of environmental services. Existing swaps also support the contention that, in developing countries at least, the constraining factor in completing ownership rights to natural resources is not the legal assignment of title but rather the degree of enforcement practiced on existing nominal claims. The fact that swaps are taking place, despite the public good nature of the services involved and inherent enforcement difficulties, suggests that the potential gains from trade may be very large.

Because all debt-for-nature swaps concluded to date are international in scope, transaction costs are compounded in many ways. Third-party enforcement of an agreement is hampered if one of the parties is a sovereign state, since a national government cannot be expected to enforce provisions that are contrary to the interests of its leaders. In response, debt-for-nature swaps typically do not include host government agencies as parties with substantive responsibilities, and they limit prescribed actions to those that are consistent with the interests of host nations. When the group donating debt is an international CO, available enforcement measures are largely limited to applying political pressure and to withholding future contributions. Accordingly, remedies for default are not included in such contracts.

The enforcement problems that result from international contracts for environmental services further illustrate how free trade agreements may help be useful in encouraging more contracting. To the extent that countries depend on one another through trade, their incentive to honor contracts is strengthened. Furthermore, the NAFTA affords an opportunity for Canada, Mexico, and the United States to seek international enforcement mechanisms that will facilitate dispute resolution. If developed, such mechanisms could reduce transaction costs substantially and create an atmosphere far more conducive to debt-for-nature exchanges and other environmental agreements. This could be an important force for environmental improvement in Mexico and other developing nations.

While the number of debt-for-nature swaps completed has been growing over time, the sums involved are so small that it is unrealistic to expect them to have a significant effect on the debt of developing nations. This does not imply, however, that further swaps are either unimportant or counterproductive. The funds these swaps generate often represent very large augmentations to the resources otherwise available for environmental protection in the nations involved. To the extent that the NAFTA results in lower transaction costs for debt-for-nature swaps and other trades in environmental amenities between Canada, the United States, and Mexico, it will make capital available for other uses at the same time that it enables the NAFTA to be a positive force for environmental quality.

Notes

1. Tropical deforestation in Latin America and Asia is estimated at 40 percent of original forest cover, and estimates for Africa exceed 50 percent. See World Bank (1989, 1).

2. See Sedjo (1992, 20–21) for an informative discussion of private and public good aspects of biological resources.

3. Leffler and Rucker (1991) apply this principal to timber harvesting contracts.

4. Limiting a contract's duration often reduces the size of each party's stake in disputes that arise while the contract is in force. The tradeoff is that short duration contracts must be negotiated more often. The scope for competition in dispute resolution can be constrained by agreeing to limits in advance, as a part of the contract. Clauses that stipulate a particular process for dispute arbitration, and provisions that specify choice of law, choice of forum, and waiver of sovereign immunity are examples.

5. Barzel (1989, 35) makes the same point in slightly different terms.

6. See Simpson (1992) for further discussion of such agreements.

7. According to some descriptions, the Merck-INBio contract also grants INBio a share of the gross sales from any commercial products produced. This is the key attribute of a *share* contract, which is discussed next.

8. This description is taken from Joyce (1991).

9. There is a natural third contract form, one in which INBio or the Costa Rican government effectively hires a pharmaceutical firm to perform the research and analysis needed to determine whether the substances gathered by INBio have commercial value. This kind of agreement has not been tried to date. The costs of enforcing such a contract would consist of the effort needed to police diligence in the research effort and to prevent the pharmaceutical firm from artificially synthesizing any natural substances discovered.

10. The countries involved to date are Madagascar, Zambia, Bolivia, Costa Rica, the Dominican Republic, Ecuador, Guatemala, Jamaica, Mexico, the Philippines, and Poland.

11. These figures are from World Resources Institute (1986, 18) and UNESCO (1991, 2). Highly indebted countries made significant payments during this period, exceeding 20 percent of export earnings in some highly indebted Latin

American countries (Wilke and Ochoa, tables 2806, 2809). Overall, developing countries paid $830 billion in interest and principal between 1982 and 1988.
12. Banks often carried debt on their balance sheets at face value, despite discounted secondary market prices. In the United States, pronouncements in the Generally Accepted Accounting Principles (GAAP) and federal laws and regulations induced banks to avoid selling debt because of 'debt contamination'. This occurs when a bank sells a portion of a country's debt portfolio at a discount and then is required to mark down, to the secondary market selling price, the country's remaining debt. A bank that sold a portion of a country's debt would either have to write-down the debt on its balance sheet, which reduces reported current earnings, or provide loan loss reserves (Allocated Transfer Risk Reserve) which also reduces earnings. See Gibson and Curtis (1990, 338–40) and Sperber (1988, 392–3) for more details. A subsequent amendment to the write-down and reserve requirement allowed a bank to sell part of an LDC's debt without reducing the book value of remaining debt to that country, "provided the bank considers the remaining loans collectible" (World Resources Institute, 1992, 22). Gibson and Curtis (1990, fn 28) note that most early debt-for-nature transactions involved LDCs whose debt had already been written down, or for which loan-loss reserves had been established.
13. Two examples are illustrative. Brazilian debt fell from a price of 85 percent of face value to 40 percent over this period and Argentinean debt fell from 66 percent to 22 percent.
14. In a *New York Times* article he stated that, "under the best circumstances, debtor nations find it hard to address critical conservation problems because of multiple social needs . . . stimulating conservation while ameliorating debt would encourage progress on both fronts" (Lovejoy, 1984).
15. For example, a swap involving Ecuador in 1989 included The Nature Conservancy, World Wildlife Fund, and Missouri Botanical Garden. Two debt-for-nature swaps included Sweden, and one included Holland.
16. These include Pew Charitable Trust, Jessie Smith Noyes Foundation, Associations Ecologica La Pacifica, John D. and Catherine T. MacArthur Foundation, Swedish Society for the Conservation of Nature, W. Alton Jones Foundation, Missouri Botanical Garden and the Organization for Tropical Studies (World Resources 1992, 309).
17. With the exception of the first swap involving Bolivia, host countries have issued local currency bonds to finance the contractual conservation obligations. The bond terms range from four to twenty years, with interest from the bonds paid annually. World Resources Institute (1992, 309) reports that $61 million in 'conservation funds' have been generated to date. This figure equals the face value of local currency bonds issued by host countries. It does not include interest, nor is it in expected present value. Unless otherwise noted all dollar amounts cited are in U.S. currency.
18. The dollar value and the number of transactions per year include both

publicly and privately funded swaps, but exclude swaps under the Enterprise for the Americas Initiative, discussed next.

19. Countries participating in the EAI must establish an environmental framework agreement, which designates conservation activities. After a country's debts are reduced, the remaining principal is paid to the United States and interest is paid into a conservation fund.

20. The P.L. 480 program offers low interest, long term loans to developing countries to purchase agricultural products from the United States.

21. The following outline of the conversion process follows von Moltke and DeLong (1990).

22. Only two U.S. banks have donated debt to a U.S. CO, the Fleet National Bank of Rhode Island which donated $254,000 of Costa Rican debt to TNC in 1987, and Chase Manhattan which donated to CI approximately $400,000 of Bolivian debt in 1988. Apparently neither of these donations were motivated by tax considerations; the debt was considered to be "nuisance" debt. See Conservation International (1991, 27–8) and Gibson and Curtis (1990, 383–4).

23. Negotiated buy-backs have, in some instances, reduced outstanding debt significantly. In 1988 Mexico retired $1.38 billion in debt, at a cost of $480 million, and Bolivia spent $34 million to buy back $308 million.

24. See Deacon and Murphy (1992) for a more detailed analysis of the role of *debt* as the *quid pro quo* in debt-for-nature swaps. In addition to scrutinizing the claims that debt causes deforestation, the authors also evaluate the possibility of gains to debtor nations from reductions in debt overhang and claims that debt purchased in the secondary market provides leverage to conservation organizations.

25. The description that follows is from World Wildlife Fund (1990).

26. The area involved is the Regional Conservation Unit of Talamanca, a mountainous area that covers about 12 percent of the country's national territory. The trust account is held by the private Banco Cooperativo Costarricense, R.L. Although the preliminary 'whereas' statements may seem superfluous, they establish the agreed objective of the document and the agreed roles and intents of the parties. This narrows the range for future disagreement regarding the actions and responsibilities of the participants and may thereby lower dispute resolution costs. An Ecuadorian agreement between The Nature Conservancy and Fundacion Natura, which is similar to other debt-for-nature contracts that TNC has negotiated, contains very similar opening statements. See The Nature Conservancy (1989).

27. The stated limit for this swap was $600,000 in face value of debt; the actual amount presented by WWF was $550,488. Two other participants, the Swedish International Development Authority and The Nature Conservancy, completed separate agreements amounting to almost $4.5 million in Costa Rican debt. The face value of the Costa Rican bonds obtained equaled 100 percent of the face value of the external debt exchanged, using the government's 'official' exchange rate. The bonds pay interest at a fixed annual rate of 8 percent and

mature in 20 years.

28. An attachment requires that activities be directed toward the target area, the Regional Conservation Unit of Talamanca in this case. In case of disputes regarding its interpretation, the English version of the text is controlling in all cases except those involving legal action brought in Costa Rican courts, where the Spanish version is controlling. The document also provides that one representative of WWF shall be appointed to the government's coordinating commission for the park involved, the Comision Coordinadora Interinstitucional de la Reserva de la Biosphera de la Armistad.

29. The standard contracts for both the World Wildlife Fund (March 20, 1990) and The Nature Conservancy (March 22, 1989) use such language. Establishing and maintaining buffer zones is also among the tasks specified in a recent proposal to the World Bank's Global Environmental Facility for protection of the Peruvian Amazon. See Alderman and Munn (1992, 42–6).

30. One Ecuadorian official remarked: "It is absurd to pay the debt . . . [I]n a few years there will not even be any point in negotiating on the debt and swaps will become meaningless" (quoted by Greener, 1991, 168). One way to avoid both inflation and non-repayment risk is to swap debt instruments for cash in the host country's currency, which is then used to endow a trust. This option was followed in an agreement in the Philippines. See Hamlin (1989, 1070). It does not, however, deal with the probability that the fund will be seized.

31. Only states can bring an action before the International Court of Justice. One might expect the United States to intervene on behalf of a U.S.-based CO seeking redress, but this possibility was ruled out when the United States surrendered its capacity to bring actions before this tribunal. This resulted from a dispute in the court's handling of a case brought by Nicaragua against the United States. See Hrynik (1990, 161).

32. For additional commentary on the lack of explicit enforcement mechanisms in swap contracts, see Hamlin (1989, 1085–6); Gibson and Curtis (1990, 342); Greener (1991, 124–31, 159–62, 188–9); Hrynik (1990, 155–6); and Lachman (1989, 153).

33. In most of the swaps negotiated to date, the party seeking to enhance environmental protection is an international CO. In 1991 and 1992, however, the U.S. government negotiated three debt-for-nature swaps under the Enterprise for the Americas Initiative. The contracts for these exchanges differ in structure from those described here, primarily due to the fact that the U.S. government can exercise different enforcement mechanisms than international conservation organizations. See Deacon and Murphy (1992) for further details.

References

Alderman, Claudia L., and Charles T. Munn. 1992. In search of resources. *Ecologica*. Bogota, Colombia: Politica-Medio Ambiente-Cultura, 9 (Jan./Feb.): 42–6.

Barzel, Yoram. 1989. *Economic analysis of property rights.* Cambridge: Cambridge University Press.

Conservation International. 1991. The debt-for-nature exchange: A tool for international conservation, update. Washington, DC.

Deacon, Robert T., and Paul Murphy. 1992. The structure of an environmental transaction: The debt-for-nature swap. Department of Economics, University of California, Santa Barbara.

de Meza, David, and J. R. Gould. 1992. The social efficiency of private decisions to enforce property rights. *Journal of Political Economy* 100(3): 561–80.

Gibson, J. Eugene, and Randall K. Curtis. 1990. A debt-for-nature blueprint. *Columbia Journal of Transnational Law* 28(2): 331–412.

Greener, Laurie P. 1991. Debt-for-nature swaps in Latin American countries: The enforcement dilemma. *Connecticut Journal of International Law* 7.

Hamlin, Timothy B. 1989. Debt-for-nature swaps: A new strategy for protecting environmental interests in developing nations. *Ecology Law Quarterly* 16:1065–88.

Hrynik, Tamara J. 1990. Debt-for-nature swaps: Effective but not enforceable. *Case Western Reserve Journal of International Law* 22:141–63.

Joyce, Christopher. 1991. Prospectors for tropical medicines. *New Scientist* 19 (October): 36–40.

Lachman, Marianne. 1989. Debt-for-nature swaps: A case study in transactional negotiation. *Journal of Contemporary Legal Issues* 2.

Leffler, Keith B., and Randal R. Rucker. 1991. Transactions costs and the efficient organization of production: A study of timber-harvesting contracts. *Journal of Political Economy* 99(5): 1060–87.

Lovejoy, Thomas. 1984. Aid debtor nations' ecology. *New York Times* (Oct. 4): A31, col. 1.

The Nature Conservancy. 1989. Debt-for-nature agreement between The Nature Conservancy and Fundacion Natura. Arlington, VA, March 22.

Patterson, Alan. 1990. Debt-for-nature swaps and the need for alternatives. *Environment* 32 (December).

Reilly, William K. 1990. Debt-for-nature swaps: The time has come. *International Environmental Affairs* 2(2): 134–40.

Sanchez, J. Angela. 1992. More money and less conditions. *EcoLogica.* Bogota, Colombia: Politica-Medio Ambiente-Cultura, 9 (Jan./Feb.): 54–7.

Sedjo, Roger A. 1992. Property rights, genetic resources, and biotechnological change. *Journal of Law and Economics* 35 (April): 199–213.

Simpson, David R. 1992. Transactional arrangements and the commercialization of tropical biodiversity, working paper. Resources for the Future, Washington, DC.

Sperber, Sebastian. 1988. Debt-equity swapping: Reconsidering accounting guidelines. *Columbia Journal of Transnational Law* 26(2): 377–406.

UNESCO. 1991. Debt for nature, *Environmental Brief*, no. 1.

von Moltke, Konrad, and Paul J. DeLong. 1990. Negotiating in the global arena: Debt-for-nature swaps. *Resolve,* no. 22. The Conservation Foundation.

Wilke, James W., and Enrique Ochoa, eds. 1987. *Statistical abstract of Latin America.* Latin America Center Publications, University of California, Los Angeles.

World Bank. 1989. *Government policies and deforestation in Brazil's Amazon region.*

World Resources Institute. 1986. *World resources 1986, A report by the World Resources Institute,* in collaboration with The United Nations Environmental Programme and The United Nations Development Programme. New York: Oxford University Press.

————. 1992. *World Resources 1992–93, A report by the World Resources Institute,* in collaboration with The United Nations Environmental Programme and The United Nations Development Programme. New York: Oxford University Press.

World Wildlife Fund. 1990. Debt-for-nature agreement between World Wildlife Fund, Inc., and Ministerio de Recursos Naturales, Energia y Minas, and Fundacion de Parques Nacionales. Washington, DC, March 20.

7

Bootleggers and Baptists—
Environmentalists and Protectionists:
Old Reasons for New Coalitions

Bruce Yandle

Introduction

The Prospect and Challenge of Free Trade with Mexico

Few things offer brighter prospects for prosperity in our times than the vision of an open market in the Western hemisphere. Think what is at stake in the pending Canada–Mexico–U.S. free trade agreement: A market that extends from the Klondike to the Yucatan, one that could tightly link the United States to its number one and two trade partners, a single trading region that could connect more than 350 million consumers.[1]

Market size is one thing, but the prospects for growth are something else to consider. Mexico's youthful population provides two million new workers annually. By contrast, the United States, with three times the population, produces one million work age people each year. If the additional workers can be matched to employment opportunities brightened by free trade, the prospects for future income growth look good indeed.

In addition, an agreement linking Mexico to her northern neighbors would be a major step in the direction of forming a single market—North and Latin America where a population of 700 million, larger than all of Europe and the former U.S.S.R. combined, could be linked together through open markets and free trade.

Consider the income gap to be closed and raised across the three major trading partners. Mexico has a per capita GNP of $2,000; Canada, $18,000 and the United States, $19,840. Experience teaches that integrated markets raise incomes across the territory, and that the lower income country would likely benefit most from the adjustment. All people taken together will gain. Always, the ordeal of change means that some people will suffer.

The vast amount of human capital to be guided by specialization is joined by massive reserves of natural resources, access to renowned harbors, and some of the world's best universities. The region has an agricultural base that is capable of feeding a doubling or tripling of the population. Though markedly different in their origin, the nations in the hemisphere share legal traditions that in principle recognize the right to contract and importance of property rights.

Adam Smith would be proud, but while Smith and his followers would celebrate the brighter human prospect offered by the simple act of allowing free people to engage in voluntary exchange across political boundaries, the inescapable fact of political boundaries means that opening doors to trade will not come easily. Rents obtained and protected through sustained and massive political efforts are at stake. Income losses that might befall highly vocal, easily identified special interest groups weigh in one side of the scale of freedom. Welfare gains that could accrue to millions of unidentifiable people weigh in the other side. Weighty environmental issues lie beyond these concerns about income and wealth.

Theory and common sense tell us the benefits of freedom far outweigh the value of artificial rents to be earned from restricting the free spirit of man. Yet another body of theory and logic tells us that an organized few can readily subvert the hopes and efforts of the unorganized millions. Freedom always brings uncertainty as unknowable outcomes wait in the wings. The trade negotiators, however, deal with the known interest group concerns.

This chapter examines the negotiations for the North American Free Trade Agreement (NAFTA) that began formally in June 1991 when trade ministers from Canada, Mexico and the United States met in Toronto. Concerns that entered the negotiations are discussed in the light of an interest group struggle that has joined together two strong special interest groups to oppose outright free trade. Modern day mercantilists who favor protection from foreign competition form the first group. The opposition they bring to the struggle is as old as nationhood and trade. They simply want political protection from competition. The second group's participation in a free trade struggle is decidedly new. Their opposition to open borders between the United States and Mexico is based on concern for environmental protection.

The story to be related here tells us the environmentalists have achieved astounding success. For the first time in history, environmental regulation rests on the trade negotiators' table as a major element to consider when attempting to whittle down barriers that prevent ordinary people in both countries from engaging in voluntary exchange.

The chapter is organized as follows. The next section provides more detail on the opposition that formed to question the logic of the NAFTA. Calling on the theme of "Bootleggers and Baptists," the discussion there explains how the two groups can achieve a mutually beneficial result.[2] The third section takes a closer look at the environmental issues and discusses alternatives that could be chosen to address those issues. The alternative chosen is then discussed in the light of the special interest theory described earlier. The fourth section describes the environmentalists' political victory and tells how the political mechanism responded to their demands. A final section summarizes the story and offers final thoughts.

Bootleggers, Baptists, and Free Trade

Players in the Special Interest Struggle

At first glance, it would seem relatively easy to identify groups that would quickly oppose expansion of trade between the United States and Mexico. The list begins with workers in protected industries. There are United States labor unions that seek to reverse a long sustained loss of members. There are owners of protected capital in United States industries that hope to delay yet another competitive struggle to hold on to threatened U.S. markets. Added to these are United States fruit and vegetable growers who fret about added competition from Mexico, which already serves one-third of the United States market, and raise concerns about pesticides and food safety. As always, there are politicians at all levels who seek to serve the interests that sustain them. Generally, those interests prefer the status quo. If change is to come, they argue, let it come slowly and painlessly.

But while these predictions are valid, more is discovered by probing deeper into the issues. It turns out that U.S. tariffs on Mexican imports are decidedly low, falling into the range of 3 to 4 percent for most items, but reaching 22 percent for copper and 10 percent for many steel products.[3] Tariffs and duties tell only part of the story. Quotas or "voluntary restraint agreements" are often more binding on trade than the duties they complement.

At present, steel and textiles are subject to such quotas. But the textile quota has been modified to allow Mexico to expand exports of goods produced with United States made fabric.[4] As it turns out, production in Mexico from U.S. fabric is preferred to the alternative—production in Asia from fabric produced in that region. Marketing orders on agricultural and horticultural products that set quotas for United States markets must also be considered. These currently affect Mexican exports and if relaxed would lead to a significant expansion of shipments to the United States. On the other side of the border, U.S. producers of processed and canned foods would enjoy an increase in exports to Mexico.[5]

Then, there are restrictions that affect U.S. investment in Mexico that will be relaxed in a true free trade environment. Limitations on foreign ownership of plants, restrictions on domestic content, and outright prohibitions on the importation of refined petroleum products pose high hurdles for United States firms that seek to gain the advantages offered by Mexico's market.

A review of the likely effects of a NAFTA on major manufactured products by the U.S. International Trade Commission indicates exports to the United States and increased United States investment in Mexico could be affected significantly for automotive products, petrochemicals, and glassware (see U.S. International Trade Commission 1991, x–xvii). These effects result from reductions in tariffs and changes in Mexico's restrictions on foreign investment. The Commission notes that while the effects could be significant, the Commission report suggests the resulting dislocations would be minimal.

Nonetheless, labor unions and management of threatened industries join ranks and argue in terms of level playing fields and fair competition. They point to low wages in Mexico and argue that the same fate awaits American workers if they are forced to compete in open markets.[6] Too much focus on the current picture may cause one to miss a major point in the protectionists' argument. The longer run outlook says that economic growth in Mexico will accelerate giving a brighter future for firms in that region than for similar ones in the United States. Workers with specialized training and a desire to maintain a way of the life in the United States recognize a threat when they see one.

Turning to broader public interest concerns, spokesmen for labor unions express concern for environmental quality and raise cautions about the sad prospects for degradation of precious natural resources. They also argue that U.S. food standards will not be satisfied, forgetting that the FDA already controls the quality of all edible products brought into the United States.

The Office of the U.S. Trade Representative (1991) has addressed the environmental issues. Its analysis of environmental concerns identified U.S. industries that have the highest compliance costs for meeting U.S. environmental regulations and then matched that list with industries that benefitted most from tariffs and quotas. The result of the analysis identified eleven industries that are vulnerable to the twin effects of environmental rules and the expected relaxation of tariffs and investment restrictions.

The industries include specialty steel, petroleum refining, five categories of chemicals including medicinal compounds, iron foundries, blast furnaces and steel mills, explosives, and mineral wool (Office of the U.S. Trade Representative 1991, 141). The Commission's report notes that the eleven industries have high capital intensity, which they point out reduces the likelihood that plant owners would take an early write off of capital in order to take advantage of lower environmental costs in Mexico. The fact that excess world capacity exists for some of the "threatened" industries also reduces the validity of arguments about relocating plants. However, world excess capacity does translate into greater effort to maintain protection from new foreign competition.

Lacking the ability to turn the tide on the basis of apparently self-serving arguments, those who seek to hold on to their rents welcome as allies all who seek to limit trade on any plausible basis. As hinted at in these remarks, there are bootleggers who welcome the support of Baptists in the struggle to restrict commercial activity.

The Mexican Standoff

The Bootlegger–Baptist analogy is a familiar one in many parts of America. Southerners, for example, have a long tradition of limiting the sale—but not the consumption—of alcoholic beverages on the Sabbath day. Baptists, Methodists, and other strong Protestant groups lobby hard to limit access to demon rum. Those who know the story, recognize another group that quietly celebrate every successful attempt to close down the legitimate seller of booze. The bootleggers—the illegal operators who take over the market when the legal seller pulls his blinds. Bootleggers and Baptists work the politicians separately but toward the same result. The bootleggers never have to march on state capital buildings calling for an end to the sale of beer, wine and booze on Sunday. They don't have to. The Baptists fight the open battles for them.

It is good that the Baptists are there, at least from the standpoint of bootleggers. The Baptists make a moral appeal that yanks at the heartstrings. They lift the debate to higher ground. What politician would openly argue that alcohol is a good thing? Of course, the bootleggers conveniently make contributions to fund the campaigns of politicians who support the Baptists' cause, which is their own.

In describing special interest players, the traditional foes of free international trade were described in some detail. What about the Baptists? Who are they? How well are they organized? What are their arguments?

The environmentalists are chief among the Baptists in the NAFTA debate. They have real concerns to be addressed. If successful, can they truly assist the American bootleggers? That is, will the methods chosen to address environmental concerns strengthen the cause of labor unions and modern day mercantilists who argue for level playing fields? Or will the methods serve the cause of free market competition?

The Limits of the Bargaining Process

There are limits to the assistance that environmental groups can provide U.S. special interests who seek protected markets. If environmental rules imposed on Mexico are too strict, U.S. firms that rely on Mexican production of component parts but seek a protected market for finished goods will suffer. If the environmental rules are too lax, they may not offset the reduction in tariffs that emerge from the negotiations. There is balance to be struck, and maintaining the status quo on trade flows probably describes the goal of the protectionists who want some environmental restrictions.

A more serious outcome threatens domestic producers if the trade negotiators wipe out tariffs and are also totally insensitive to environmental concerns. Owners of newly built production in Mexico will gain two advantages over their domestic counterparts who face higher environmental compliance costs. Their production costs will be relatively low, and there will be no border tax to pay when final goods enter the United States market.

The problem can be understood quickly by visualizing a demand curve for steel in the U.S. market and the associated aggregate supply curve that includes the supply curve of U.S. produced steel and a supply curve for steel imported from Mexico. There are tariffs and quotas that currently limit the quantity of steel shipped from Mexico. The U.S. price is raised by the restriction. Domestic firms and labor are protected.

If the trade negotiators successfully reduce the tariffs and quotas and pay no heed to environmentalists, the U.S. price will fall. Protection will end. But if sufficient environmentally based restrictions are substituted for the tariffs and quotas, protection will continue, even in a free trade environment. To assist the protectionists, the environmentalists must push for particular kinds of controls, such as industry standards that limit output. In addition, they can argue for one-time environmental impact fees that will blunt the attractiveness of Mexican investment opportunities.

In short, the most costly forms of regulation will best serve the combined interests of environmentalists and protectionists. Protection of environmental assets based on property rights, performance standards, and emission fees will not deliver an outcome that fixes quantity on an industry-by-industry basis. Instead, these alternate instruments deliver environmental protection but allow market forces to determine the ultimate mix of goods and services produced in the trading economies. In other words, American style command-and-control that specifies precise methods of control on an industry-by-industry basis is the preferred outcome of the special interests. Command-and-control is the environmental counterpart of tariffs and quotas.

The Environmental Issues and the Free Trade Alternative

Alternatives for Addressing Real Problems

The environmentalists do not have to search hard to find a basis for their concerns because there are obvious environmental problems in Mexico, just as there are in many developing and developed countries. Mexico's problems spring significantly from the huge population concentrated in Mexico City. Claiming title to the second largest urban center in the world, Mexico City holds 20,000,000 people, almost one-fourth of the country's population, with a population density of 37,000 per square mile, which is more than twice that of Los Angeles.[7] Some 1,500 more arrive each day.

There are 40,000 cabs plying the trade in Mexico City and hundreds of thousands of automobiles and trucks operating in the congested streets (see U.S. Congress 1991, 4). Many of the vehicles are old. Most lack emission controls. And the low quality of the gasoline burned generates heavy emissions.

Movement to Mexico City has become a way of life for young people from the rural regions of the country. Until recently, there have been few opportunities elsewhere that could match Mexico City's combination of welfare and employment as domestic help. Things are different now. Industrial development is spreading in northern Mexico, and the central government is taking steps to reduce the subsidies that misguide people as they make location decisions.

But the heavy concentration of older manufacturing facilities in Mexico City and a few other locations, and the understandable near absence of systematic approaches for dealing with environmental scarcity compounds the problem. But the environmental problem is far more than one of physics or biology. It is economic. Extremely low incomes and a skewed distribution of income contribute fundamentally to the problem. Simply put, extremely poor people who can hardly form a vision of the future, let alone prepare for it, cannot be expected to push for tight pollution standards that will generate benefits far into the future. The fact that most of Mexico is sparsely settled desert terrain adds another complicating dimension to the environmental control problem. Most of the country has little cause for environmental concerns.

As one might expect, the Mexican people are beginning to address the major environmental problems found where population concentrations are high. Like their U.S. counterparts, Mexican regulators have moved toward a system of command-and-control that imposes standards, closes plants, and restricts voluntary use of automobiles. Indeed, Mexico's General Law for Ecological Equilibrium and Environmental Protection is said to be patterned after "U.S. law and experience."[8]

Since 1988, the Mexican government has imposed 82 permanent plant closings and almost a thousand temporary shutdowns for environmental reasons. But the rate of industrial growth in a few key areas is outstripping the ability of regulators to impose controls. As in some regions of the United States, the demand for higher incomes in Mexico far outweighs the political demand for restricted growth. At this point in the country's development, an excess supply of environmental quality is an asset to be used. Conservation predictably will arrive later when incomes are higher. That day will arrive sooner if free trade expands.

Problems with the Maquiladora Region?

While Mexico City offers the most challenging environmental problem for the citizens of Mexico, environmentalists who have joined the battle against the NAFTA focus their attention on the maquiladora region, which is a 250,000 square mile area located in a 20-kilometer strip along the United States–Mexico border.[9] Established in 1965 as a free trade zone that allows producers to bring in component parts duty-free and manufacture goods for export to other parts of the world, the strip now contains close to 2,000 manufacturing plants, mostly U.S. owned. Nearly 80 percent of Mexico's exports of manufactured goods to the United States originate in the maquiladora (see Jenner 1991, 37). Some 400,000 workers are employed, mostly women who earn low wages.

The firms located in the region pay minimum taxes and meet lax environmental standards. By law, all toxic and hazardous wastes produced in the manufacturing plants and not processed according to Mexican standards must be returned to the firms' country of origin, but enforcement of the law is far from perfect. According to environmentalists, most of the plants pose severe health and environmental risks to workers and nearby communities (see National Wildlife Federation 1991). Some five million people live in the maquiladora region, and the number is growing.

Viewed in terms of wealth production, the maquiladora experiment is an astounding economic success. In 1965 there were 12 plants in the region employing 3,000 people, as compared to today's 2,000 plants employing 400,000 workers. However, when viewed as an activity that uses environmental resources while improving human well being, the success of the program is questioned by environmentalists.

Their concern is not without justification. According to a review of U.S.–Mexico environmental issues conducted by the Office of the U.S. Trade Representative (1991, 4), emissions of carbon dioxide and ozone levels exceed U.S. standards in El Paso/Jarez and San Diego/Tijuana. There are similar water quality and municipal waste problems. Aquifers that provide groundwater for wells are threatened; the quality of rivers along the border is deteriorating. The level of other pollutants is far above the U.S. standards in other border counties. But questions posed on the basis of U.S. standards must be laid to rest when alternatives are considered. The important question to ask is "Relative to what?" What if the free trade zone had not been formed? Where would the four million people be living and in what condition?

The report of the Office of the U.S. Trade Representative (1991, 4) asked the "relative to what" questions. Making comparisons of expanded pollution with and without a free trade agreement, the report concludes that pollution will grow in either case. However, without the trade agreement and the gains that come with it, emission growth will range from plus 40 percent to plus 225 percent in the next 10 years. With the agreement, pollution will expand from zero to 165 percent in the same decade.

These estimates notwithstanding, some environmentalists argue the maquiladora program will grow with the NAFTA. They forget that the program is stimulated by tariff forgiveness. If tariffs are reduced extensively, incentives to locate in free trade zones will be blunted. Indeed, all of Mexico could be a free trade zone. Instead of concentrating environmental use in one location, activities will be spread. While the environment will be used, conservation problems will be less concentrated.

While the pressures of a large population inadequately served by sewage treatment and water purification form serious environmental problems, other pieces of the border problem come from rules and regulations that could be reduced by the NAFTA. Regulations that relate to the transportation of goods into and from Mexico require the use of Mexican drivers and trucks when operating south of the border. The border has the heaviest traffic of any in the world. In a typical year, 200 million people will pass through one of the major portals (Office of the U.S. Trade Representative 1991, 70).

Inspections of merchandise and paperwork form another delay at the congested portals connecting the two countries. Taken together, the combined border and transportation controls generate massive traffic jams, where a thousand vehicle operators idle their engines as they wait in line. The resulting severe emission loadings could be reduced by simplified rules under the NAFTA.

Current trade restrictions that limit the importation of natural gas from the United States to Mexico cause Mexican industry to burn higher polluting petroleum products. Limits on foreign investment mean that older, more polluting, plants operate longer than they would otherwise.

But the unregulated expansion of foreign investment in Mexico is another environmentalist concern. Kodak is setting up shop in Monterrey. Other electronic and chemical firms are expanding at other Mexican locations. The newly entering U.S. firms are described as fleeing tight United States environmental standards. In some cases, the allegation is probably correct.

For example, in 1988 when California's South Coast Air Quality Management District promulgated strict rules for controlling emissions from spray paint operations, some 40 affected firms relocated to the Mexican border, where standards were lax (see Ganster 1991, 16). Putting anecdotal evidence to one side, the savings that accrue from lax environmental standards are generally too small to justify a Mexican move for that reason alone. To counter the environmental runaways, the environmentalists still call for Mexican environmental standards and controls that compare with U.S. standards.

The concerns do not end with problems that emerge when firms locate plants in Mexico. They reach out and address America's "addiction" to fossil fuels and how that unfortunate habit will spread south of the border. Mexico's petroleum supplies are second only to those of Saudi Arabia and would offer North America a more dependable energy source, which would postpone efforts to break the fossil fuel habit. Mexico's state-owned petroleum industry is hungry for new capital. Expanded trade will surely lead to an expanded production of petroleum. Evidence from the Canadian–United States trade expansion is persuasive. Since that agreement fell into place, Canada has increased her production of timber products, and the increased demand from exporting industries has led to increased production of electricity and the related consumption of fossil fuels.

Other more complicated concerns are raised by environmental groups. Environmentalists worry about the push to require Mexico to recognize intellectual property rights, especially as they relate to pharmaceutical product development. Their argument is that recognition of U.S. patents and copyrights will cause U.S. companies to gain profits by producing new biological products. These profitable ventures will increase the use of Mexico's natural biological resources that provide inputs to the production of new medicines. However, environmentalists contend that without special protection, Mexican resources will be sold as commodities, while the patented end products will earn a handsome return for U.S. corporations.

Free Trade Threatens Meaningless U.S. Regulations

While global competition inevitably grinds down inefficient production of goods and services forcing producers worldwide to provide more value from their use of scarce resources, the same grinding effect works on producers of inefficient regulation. As a part of trade negotiations, all parties could be challenged to show that their environmental regulations are based on scientific evidence and are necessary to protect the public's interest in health and environmental quality (Wirth pers. comm. 1991). Regulations shown to be invalid on the basis of solid scientific evidence could be declared nontariff barriers and as such, negotiated away. U.S. regulators and their supporters could be placed in the position of the emperor, whose new clothes turned out to be sheer imagery. As the National Wildlife Federation's collective statement on the topic put it:

> The ability of policy makers to promote natural resource conservation through trade policies, including import and export bans, environmental levies, and labeling requirements, can also be severely limited. Since the implementation of the U.S.–Canada Free Trade Agreement . . . , British Columbia has been forced to eliminate a provincially funded tree planting program which American timber interests considered to be unfair. (National Wildlife Federation 1990, 10)

But the beneficial prospects brought by international competition in regulation may have been cut off at the pass. Bowing to environmentalists' pressures, President Bush has indicated that trade negotiators will not weaken U.S. environmental and health laws as part of the final agreement. As the President's carefully worded statement put it: "The United States will maintain the integrity of the U.S. regulatory process, which is based on available scientific evidence" (U.S. Congress 1991, 9). If the President is correct in asserting that U.S. policy is based on scientific evidence, as opposed to politics, there is significant reason for doubt about the adequacy of the evidence.

The Environmental Solution

The Cartel Net Tightens

Pressure brought to bear on U.S. government officials has generated an astounding success for the invisible coalition of bootleggers and Baptists. Congressional concerns have been raised. The EPA has responded with major initiatives that will bring U.S. environmental control to bear on Mexico's industry and population, and even more far-reaching proposals are on the table. Recognizing command-and-control regulation as a cartel-forming device that restricts output, raises prices, and protects competitors from fringe-firm competition is one of the major insights provided by regulatory economists (see Buchanan and Tullock 1975). Use of economic incentives, such as property rights and common law rules that force all parties to recognize and pay for use of the environment cannot accomplish the same end. For this reason, command-and-control is supported reluctantly by industry, and the same instrument is pushed by environmental organizations and regulators. The regulators seek to expand their domain of authority. Environmentalists mistakenly see command-and-control as providing certainty.

The environmentalists are calling for border taxes—not tariffs, but taxes that will be used to fund costly environmental cleanup. What is gained in the way of tariff reductions on the one hand could be lost through the imposition of environmental taxes on the other. Barring success in that direction, environmentalists have called for a tax on new capital investment that will represent payment for the remediation of environmental problems. Of course, the funds for environmental control and cleanup will have to come from somewhere, but the likelihood that the funds will be spent in ways that promote efficient and effective environmental control is dim. A tax-fed slush fund for remediation and command-and-control regulation places power in the hands of bureaucrats who have no reliable incentive to minimize cost.

The U.S. EPA has gained legislative authority to embark on regulating the United States–Mexico border, and is also working with the World Bank to develop standards and controls to be applied in Mexican industry. The Mexican government has already moved to close plants as a way of reducing pollution. The EPA and the Mexican regulatory agency, SEDUE, are cooperating in training and enforcement programs. The result of all this may lead to cleaner air and water, but it will surely lead to higher production costs, fewer jobs, and selective enforcement that favors existing firms and established special interests.

Summary and Conclusion

Regardless of the outcome of the NAFTA legislation, one can safely bet that trade between Canada, Mexico, and the United States will expand. Something of substance will emerge from the negotiations. The process for trade expansion has been underway for too long; the pressure for change is simply too great. Mexico is emerging as an industrialized nation; the free trade zone experiment has proven that the prospects for mutually beneficial trade are large. When the final agreement is reached, we also can be certain that environmental concerns will be addressed. The concerns are real; the environment is at risk; too much is at stake to dismiss the concerns out-of-hand.

But the combined forces of modern day mercantilists and environmentalists will likely yield an outcome that leaves untapped welfare gains on the table. While trade expands and incomes rise, the newly imposed environmental regime will bring a new layer of high cost bureaucratic controls. Environmental controls will partially replace trade barriers that previously limited trade. Efficiency gains on one side of the ledger will be partly, if not completely, offset by efficiency losses on the other side.

Notes

1. Free trade agreements (FTAs) are negotiated under rules established by members of GATT, the General Agreement on Tariffs and Trade, even though parties to FTAs generally develop preferences for trade with each other, which is contrary to the purpose of GATT. Under the rules, the FTA must cover virtually all trade; duties must be eliminated in a timely fashion; and duties imposed on nations outside the agreement must be no higher than former duties that affect trade between the negotiating parties. For more discussion, see Office of the U.S. Trade Representative (1991, 52).
2. For discussion of the theory, see Yandle (1989).
3. For details, see Office of the U.S. Trade Representative (1991, 138–39).
4. For discussion of the various restraints, see U.S. International Trade Commission (1990, 2–5).
5. For discussion, see U.S. International Trade Commission (1991, 4–8).
6. See U.S. Congress (1990, 220–30). In his comment, Mark A. Anderson described the low wages paid Mexican workers as being below the subsistence level. As an example, he told of workers in a Zenith plant who receive $26.16 for a 48-hour work week, providing a paycheck as evidence, but he said nothing about the workers' alternatives for employment or pay.
7. For more detail, see Office of the U.S. Trade Representative (1991, 179).
8. For a brief comment on the statute, see U.S. Congress (1991, 3).
9. A summary document distributed by the National Wildlife Federation (1991) provides a composite view of leading environmental organizations regarding the maquiladora region.

References

Buchanan, James M., and Gordon Tullock. 1975. Polluters' 'profit' and political response: Direct control versus taxes. *American Economic Review* 65 (March): 139–47.

Ganster, Paul. 1991. The two Californias. In *Free trade and the United States–Mexico borderlands: A regional report by Baylor University.* Washington, DC: Joint Economic Committee, U.S. Congress, July 1.

Jenner, Stephen R. 1991. Manufacturing. In *Free trade and the United States–Mexico borderlands: A regional report by Baylor University.* Washington, DC: Joint Economic Committee, U.S. Congress, July 1.

National Wildlife Federation. 1990. Environmental concerns related to a United States–Mexico–Canada free trade agreement. Washington, DC, November 27.

————. 1991. Trade and the environment: Information packet. Washington, DC, November 1.

Office of the U.S. Trade Representative, Interagency Task Force. 1991. *Review of U.S.–Mexico environmental issues*, draft. Washington, DC, October.

U.S. Congress. House. 1990. *Hearings before Subcommittee on Trade of the Committee on Ways and Means.* Testimony of Mark A. Anderson (AFL-CIO), Steve Beckman (UAW), and Arthur Gundersheim (Amalgamated Clothing and Textile Workers Union), June 14, 28.

U.S. Congress. 1991. *Response of the administration to issues raised in connection with the negotiation of a North American free trade agreement.* Transmitted to the Congress by the President, May 1.

U.S. International Trade Commission. 1990. *Review of trade and investment liberalization measures by Mexico and prospects for future United States–Mexican relations.* Washington, DC: International Trade Commission, April.

————. 1991. *The likely impact on the United States of a free trade agreement with Mexico*, USITC Publication 2353. Washington, DC: U.S. International Trade Commission, February.

Wirth, Timothy E. (Senator). 1991. Letter to William K. Reilly, EPA Administrator, May 7.

Yandle, Bruce. 1989. *The political limits of environmental quality regulation.* Westport, CT: Quorum Publishers.

About the Authors

Terry L. Anderson, PERC Senior Associate, is Professor of Economics at Montana State University. He is co-author with Donald Leal of *Free Market Environmentalism* (Pacific Research Institute and Westview Press), author of *Water Crisis: Ending the Policy Drought* (Johns Hopkins Press), and editor of *Water Rights* (Pacific Institute for Public Policy Research). He is editor of *Property Rights and Indian Economies* (Rowman & Littlefield Publishers, 1992), and co-editor with Randy Simmons of the forthcoming volume, *The Political Economy of Customs and Culture: Informal Solutions to the Commons Problem.* In 1988, he was a Visiting Professor at Clemson University and a Fulbright Scholar at Canterbury University in New Zealand.

Robert A. Collinge is Assistant Professor of Economics at the University of Texas at San Antonio. His previous experience includes positions as an economist with the American Petroleum Institute and with the U.S. International Trade Commission, as well as a position on the economics faculty at the University of Louisville. Dr. Collinge received his Ph.D. in economics from the University of Maryland. He has authored a variety of articles dealing with international and domestic public policy issues.

Robert T. Deacon is Professor of Economics at the University of California, Santa Barbara. He received a Ph.D. from the University of Washington and has been a Fellow at Resources for the Future, Washington, D.C., and the Hoover Institution at Stanford University. His current research interests focus on tropical deforestation and forest use in developing countries. In addition, Dr. Deacon is an occasional consultant to The World Bank, the U.S. Environmental Protection Agency, and the U.S. petroleum industry.

Peter M. Emerson is Senior Economist at the Environmental Defense Fund in Austin, Texas. Prior to joining EDF, he was Vice President for Resource

Planning and Economics at The Wilderness Society. Other positions include that of principle analyst for the U.S. Congressional Budget Office, agricultural economist at the U.S. Department of Agriculture, and teacher at the University of Maryland. Dr. Emerson received his Ph.D. from Purdue University. He has been active in policy forums as a consultant.

Steven Globerman is Professor of Economics at Simon Fraser University in Vancouver, British Columbia. He holds a Ph.D. in economics with specialties in industrial organization and international business. He has published over 60 articles and 20 books and monographs on these subjects. He has also consulted for numerous private and public sector organizations on issues relating to international trade and investment, regulation and anti-trust.

Paul Murphy is a Ph.D. student and a teaching and research assistant at the University of California, Santa Barbara. He received a B.A. and M.A. in economics from California State University, Fullerton. His current research involves the political economy of tropical deforestation and resource use. At the 1993 ASSA meetings, he presented a paper (with Robert T. Deacon) entitled, "The Structure of an Environmental Agreement: The Debt-for-Nature Swap."

J. H. Patterson is Director of International and Government Relations for Ducks Unlimited Canada. He received a Ph.D. from Carleton University. Dr. Patterson has served as Director of the Migratory Birds Branch and the North American Waterfowl Management Plan Implementation of the Canadian Wildlife Service. From 1988–1989, he was President of the International Association of Fish and Wildlife Agencies in Washington, D.C. He was appointed to the federal government's Sectoral Advisory Group on International Trade for the Agriculture, Food and Beverage Industries in 1991.

Roberto Salinas-León is Executive Director of the Center for Free Enterprise Research in Mexico City. He received a Ph.D. in philosophy and political theory from Purdue University. Dr. Salinas was a Visiting Fellow at the Heritage Foundation in 1989 and 1992, and has participated in various conferences and seminars on privatization, free trade, structural reform, and liberty. He is a columnist and radio commentator and advises various entrepreneurial institutions and organizations in Mexico, Canada, the United States, and Latin America.

Bruce Yandle is Alumni Professor of Economics and Legal Studies at Clemson University where his teaching and research focus on environmental economics and regulation. His Washington experience includes serving as Senior Economist on the President's Council on Wage and Price Stability and as Executive Director of the Federal Trade Commission. Dr. Yandle has written and edited ten books in economics, including three on environmental issues. His most recent books include *The Political Limits of Environmental Regulation: Tracking the Unicorn*, 1989, and *The Economic Consequences of Liability Rules*, with Roger Meiners, 1991. He and Roger Meiners are currently co-editing a volume of papers titled *Taking the Environment Seriously*.